FOUR..........
DECADES OF

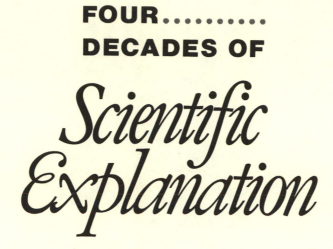

Scientific
Explanation

....

D0162024

FOUR·········
DECADES OF

Scientific Explanation

WESLEY C. SALMON
·····················

University of Minnesota Press · Minneapolis

Copyright © 1989, additional material copyright © 1990 by the
Regents of the University of Minnesota
All rights reserved. No part of this publication may be reproduced, stored in a retrieval system,
or transmitted, in any form or by any means, electronic, mechanical, photocopying, recording,
or otherwise, without the prior written permission of the publisher.

Published by the University of Minnesota Press
2037 University Avenue Southeast, Minneapolis MN 55414.
Printed in the United States of America.

Library of Congress Cataloging-in-Publication Data

Salmon, Wesley C.
Four decades of scientific explanation / Wesley C. Salmon.
p. cm.
Bibliography: p.
Includes index.
ISBN 0-8166-1825-9
1. Science—Philosophy—History. 2. Science—Methodology—History. I. Title.
Q174.8.S26 1990
501—dc20
89-5155
CIP

The University of Minnesota is
an equal-opportunity educator and employer.

In fond memory of
Cynthia A. Schuster
1910–1983
Philosopher, mentor, and true friend

Contents

Preface

The search for scientific understanding of the world has roots reaching far back into antiquity, long before Aristotle. Although it is not nearly complete, this *scientific* quest has been remarkably successful across a wide variety of fields of investigation. Philosophers and scientists, since at least the time of Aristotle, have tried to say in what such understanding consists. This *philosophical* quest has, until recently, met with considerably less success. One of my central theses is that, nevertheless, a great deal of important philosophical progress has been made in a recent period of four decades.

A little over forty years ago (1948) Carl G. Hempel and Paul Oppenheim published their epoch-making essay "Studies in the Logic of Explanation." Ironically, this now classic article was virtually ignored throughout the first decade after its appearance. Then, around 1958, it became the focus of intense discussion and heated controversy. Almost everything written on the nature of scientific explanation in the last thirty-odd years derives directly or indirectly from that essay. This book recounts these developments and offers an evaluation of the present status of the subject.

Four Decades of Scientific Explanation grew out of an NEH Institute on the topic "Is a New Consensus Emerging in Philosophy of Science?" held at the Minnesota Center for Philosophy of Science in 1985–86; the first term was devoted to scientific explanation. The way the question is posed implies that an old consensus existed—indeed, one did. It was based squarely on the 1948 Hempel-Oppenheim essay. The old consensus, alas, has not survived, and it has not yet been replaced by a new one. To assess the prospects for a new consensus, it is essential to see how we got from there to where we are today. As I suggest in the conclusion, we may now possess a basis upon which a new consensus can be forged.

Four Decades of Scientific Explanation is an introductory book; it does not presuppose any familiarity with the vast philosophical literature on scientific explanation. Starting with the most fundamental issues raised by Hempel and Oppenheim, it aims to provide an introduction to current discussions by tracing their historical antecedents. For anyone who wants to achieve a more than superficial understanding of these issues, this historical approach seems to me the best way to get into them.

Four Decades of Scientific Explanation is reprinted, without alteration, from

Scientific Explanation, volume XIII of Minnesota Studies in the Philosophy of Science (University of Minnesota Press, 1989), edited by Philip Kitcher and me. That volume contains, in addition to the present essay, papers by Nancy Cartwright, Paul Humphreys, Philip Kitcher, David Papineau, Peter Railton, Merrilee Salmon, Matti Sintonen, and James Woodward. These essays cover a wide range of topics, but one of them, "The Causal Mechanical Model of Explanation" by Woodward, is specifically directed toward my 1984 work, *Scientific Explanation and the Causal Structure of the World (SE & CSW).* Although Woodward's paper vigorously criticizes many aspects of the views set forth in that book, it is also a highly constructive discussion of the issues involved. I should like to say a little about it.

Woodward's essay raises four general questions about the causal/mechanical conception of explanation advocated in *SE & CSW :*

(1) Is it reasonable to suppose that all scientific explanations will meet the requirements of the causal/mechanical model?

(2) Does Salmon's model capture all of the features that are relevant to the assessment of scientific explanations?

(3) Should we take statistical theories like quantum mechanics as providing explanations of individual outcomes; and, relatedly, does Salmon's SR basis capture the way in which statistical evidence is relevant to the construction of explanations in QM and elsewhere in science?

(4) Should we abandon epistemic conceptions of explanation in favor of Salmon's ontic conception? Is there a single logic of explanation valid in all possible worlds or, for that matter, everywhere in our world? (Kitcher and Salmon 1989, p. 359)

The Minnesota Institute was held not long after the publication of *(SE & CSW),* and it was an extraordinary learning experience for me. This educational process continued throughout the subsequent period during which I worked on *Four Decades of Scientific Explanation.* A number of the lessons learned are mentioned in the conclusion of this essay. They reflect various shortcomings of the earlier book and have a direct bearing on the issues raised by Woodward.

For example, in *SE & CSW* I was eager to exhibit an unbridgeable chasm between the inferential and ontic conceptions of scientific explanation, and to deny that any types of explanation are arguments. As a result, I paid insufficient attention to the importance of explanations of laws and to the problems associated with such explanations. When I came to the realization—as I explain in the conclusion of *Four Decades*—that, in their present forms, the unification approach and the causal/mechanical approach are not mutually exclusive, but, rather, complementary, my attention was drawn back to explanations of laws and the relations between laws and theories. Such considerations go a long way toward answering Woodward's questions.

Consider the fourth question. In *SE & CSW* I explicitly denied that there is "any single logic of explanation valid in all possible worlds or, for that matter, everywhere in our world." I can think of *nothing* illuminating to say about scientific explanation in all possible worlds. In this world I doubt that there is any single *logic* of explanation. I believe that it is important to look for mechanisms, but I am convinced that the mechanisms of the quantum domain — which may well be noncausal — are very different from those that operate on a macroscopic scale.[1] Given that I now believe that the causal/mechanical and unification conceptions are complementary, I certainly do not maintain that "we should abandon epistemic conceptions." The fourth question receives the straightforward negative answer Woodward regards as correct.

Woodward's second question can be answered as readily as his fourth. The recognition that unifying explanations complement — rather than compete with — causal/mechanical explanations implies immediately that the causal/mechanical conception does not "capture all of the features that are relevant to the assessment of scientific explanations."

Return, then, to Woodward's first question. Again, the answer is negative for the reasons just given. However, in this context Woodward raises problems having to do with the relationship between my causal/mechanical model and explanation in the special and general theories of relativity. As a matter of fact, relativity theory played a large heuristic role in the development of the notions of causality that were set forth in *SE & CSW*. Special relativity is precisely the context in which the distinction between causal and pseudo-processes is crucial. The entire structure of the theory can be developed in terms of the behavior of light rays (actual and possible). The fundamental principles of light — that it is a first signal and that its speed is constant in any inertial system — are basic and unexplainable within the theory. The spacetime structure is given by the Minkowski light cone, which is best interpreted as the cone of causal connectibility. Given those principles, the relativity of simultaneity, length contraction, and time dilation are necessary consequences. And special relativity rules out action-at-a-distance. The foregoing explanations obviously involve derivations; they are explanations of laws that fit comfortably within the unification conception of explanation. But they also involve causal mechanisms in an essential way. Furthermore, the same causal basis can be extended to the spacetime of general relativity by the geodesic method, which adds the behavior of freely falling material particles to that of light. The causal structure of the world that is mentioned in the title of *SE & CSW* is the causal structure of special and general relativity.[2]

Another issue raised in the context of the first question concerns the applicability of the causal/mechanical account in quantum mechanics. Woodward understandably finds my discussion of explanation in that area quite confusing, for — as

I said more than once in *SE & CSW*—I am confused about it. Nevertheless, I can clear up some of the issues he raises. In the first place, as long as we see the world in terms of the four-dimensional spacetime of relativity theory, I hold out no hope for any local hidden-variable theory that would furnish common cause explanations by way of continuous causal processes for quantum phenomena. The violations of the Bell inequalities—predicted by quantum theory and confirmed by experiment[3]—seem to me to rule out that possibility entirely. But superstring theory—mentioned by Woodward as an explanatory theory that excites many physicists—cannot live in that world; it requires a spacetime manifold of higher dimension.[4] Changing the dimensionality of spacetime would present the possibility of restoring continuity where it seems now to be absent. I am *not* claiming, however, that this will be the ultimate solution to our problems about quantum-mechanical explanation.

My basic feeling about explanation in the quantum realm is that it will involve mechanisms, but mechanisms that are quite different from those that seem to work in the macrocosm. Perhaps the crucial mechanism is simply the so-called *collapse of the wave function.* But I think we must admit, in all candor, that we do not understand that mechanism and that significant numbers of physicists share this view. If that were not so, would the *many-worlds interpretation* of quantum mechanics be taken seriously? Would observation by a conscious observer be considered an indispensable aspect of microphysical explanation? Would serious physicists—e.g., David Mermin—say that anyone who is not worried about these problems has rocks in their head? Since, in company with most thoughtful philosophers, I reject the view that scientific explanation involves reduction of the unfamiliar to the familiar, I do not believe that the key to quantum-mechanical explanation is just to get used to the peculiarities we find in that domain. Perhaps such holistic moves as appeal to nonlocal conservation principles will be required. As I said at the end of *SE & CSW*, and reiterate in the conclusion of this book, an adequate characterization of quantum-mechanical explanation is a premier challenge for contemporary philosophy of science. Woodward remarks that "more needs to be said by Salmon" regarding explanation in quantum mechanics (Kitcher and Salmon 1989, p. 361); I would simply add that more needs to be said by many others as well—especially those who specialize in the philosophy of quantum mechanics.

The first part of Woodward's third question—"Should we take statistical theories like quantum mechanics as providing explanations of individual outcomes"—raises deep philosophical perplexities. Given the serious problems that arise in connection with various attempts to provide an account—such as the I-S (inductive-statistical) or S-R (statistical-relevance) models—of statistical explanations of particulars, it is strongly tempting to deny that any such things exist, and to conclude that all statistical explanation is statistical explanation of laws—more or less along the lines of Hempel's D-S (deductive-statistical) model. Perhaps we should say, as Woodward suggests, that irreducibly statistical individual events cannot be explained. We can understand how the world works, in terms of stochastic processes and mechanisms, but we cannot explain *what happens*. We can, at best, assign probabilities to the various *kinds of happenings*.

I considered this question in *SE & CSW* and came out with a negative conclusion. In a subsequent paper, "Deductivism Visited and Revisited" (W. Salmon 1988) I reconsidered it more carefully, but, again, a negative answer resulted. A negative answer is also given in this book (see §4.9). Three points seem to me to bear critically upon this issue. First, there are probabilistic phenomena — such as the melting of an ice cube in a glass of tepid water — in which the probabilities are so high that, for all practical purposes, they amount to certainties. We seem to feel that we can explain individual occurrences of this sort. The question then becomes at what level of probability we cease to be able to have such explanations. Any answer seems arbitrary.

Second, there are practical situations — cases of applied science — in which we try to explain individual events, perhaps to assign responsibility. Among soldiers who witnessed nuclear explosions at close range it was a matter of chance that a small number subsequently contracted leukemia and that one rather than another became a victim. Nevertheless, I am strongly inclined to believe that witnessing the blast caused the leukemia and that the leukemia is explained by the exposure to radiation.

Third, there is a tension between concepts of causality and concepts of explanation; it can most easily be seen in connection with J. L. Mackie's hypothetical candy machines (see pp. 176–77 below). Mackie describes a genuinely indeterministic machine in which depositing a shilling is a necessary, but not sufficient, condition of receiving a chocolate bar. He argues that, under the circumstances, if one receives a candy bar after inserting the coin, the insertion of the coin causes the machine to yield the candy. I would add that it explains the ejection of the candy. But, the probability of getting the candy if the coin is inserted is less than one. In pondering the issue of statistical explanation of individual occurrences, we should think of Mackie's other indeterministic machine as well. With this machine insertion of the coin is sufficient, but not necessary, to procure a candy bar. If one inserts a shilling and a candy bar emerges, it cannot be claimed that the insertion of the coin caused the chocolate bar to emerge, for the machine might have yielded it spontaneously. But the probability of getting a candy bar upon insertion of the coin is one. The same points can be made, incidentally, in connection with real physical detecting devices rather than fictitious candy machines. It seems to me that the relationships among causality, explanation, and probability in such situations deserve careful consideration. It may turn out, after all, that we have explanations of individual events whose probabilities are intrinsically smaller than unity. This issue has paramount philosophical importance.

Woodward also raises certain questions about my choice of examples of statistical explanation. Some of these examples came from concerns I felt deeply in the early 1970s. Hempel's doctrine of essential epistemic relativization of inductive statistical explanation raised the question of whether one could make sense of the notion of an objectively homogeneous reference class, and his high probability requirement led to problems about the explanation of events that are intrinsically improbable. It seemed heuristically helpful to deal with such examples as spontaneous radioactive decay. There was, of course, no thought that such examples qualify as serious quantum mechanical explanations; Railton's explanation of the alpha decay of a uranium nucleus comes closer precisely because it does appeal to basic laws to explain the mechanism of barrier penetration. But the simple examples I used were intended to illustrate the continuum of cases of ineluctable probability from extremely high to almost vanishingly small.

I had hoped to make it clear that considerations relating to probabilistic explanation should have interest for those working in fields other than atomic physics. For this reason, Hempel's biomedical example (strep infection) and Greeno's sociological example (car theft) were also included. One can only agree with Woodward that these examples differ fundamentally from the foregoing quantum mechanical types — especially with regard to such factors as their incompleteness, the possibility that they are at bottom deterministic, and the unavailability of high level explanatory theories. These are points that I have often reiterated.

I should like to express my sincere thanks to Jim Woodward for his stimulating and insightful critical discussion of *SE & CSW*. I am grateful to the University of Minnesota Press for making possible the publication of *Four Decades of Scientific Explanation* as a separate book. I want to thank the National Endowment for the Humanities for its support of the institute at the Minnesota Center. Finally, I should like to express my gratitude to Philip Kitcher and C. Wade Savage, co-directors of the institute, for their kind invitation to participate in an extremely stimulating workshop, and to the other participants as well for their valuable contributions.

W. C. S.
June 1989

Notes

1. I also claim, incidentally, that in certain disciplines functional explanations are legitimate, but it seems to me that they are basically causal (see §3.8).

2. Although I have, in the past, defended various conventionality theses regarding space and time, I am not at present wedded to them and they have no bearing on my claims about causal explanation.

3. As David Malament showed in a recent lecture at the Center for Philosophy of Science, University of Pittsburgh, such violations are not isolated oddities; they are pervasive features of quantum theory.

4. I do not mean to suggest that superstring theory as now understood will restore spatiotemporally continuous causal structure.

FOUR..........
DECADES OF
Scientific Explanation

· · · · · · · · · · · · · · · · ·

Introduction

The search for scientific knowledge extends far back into antiquity. At some point in that quest, at least by the time of Aristotle, philosophers recognized that a fundamental distinction should be drawn between two kinds of scientific knowledge—roughly, knowledge *that* and knowledge *why*. It is one thing to know *that* each planet periodically reverses the direction of its motion with respect to the background of fixed stars; it is quite a different matter to know *why*. Knowledge of the former type is descriptive; knowledge of the latter type is explanatory. It is explanatory knowledge that provides scientific understanding of our world.

Nevertheless, when Aristotle and many of his successors down through the centuries tried to say with some precision what constitutes scientific explanation they did not meet with great success. According to Aristotle, scientific explanations are deductive arguments; as we shall see, this idea has been extraordinarily influential. But as Aristotle clearly recognized, not all deductive arguments can qualify as explanations. Even if one accepts the idea that explanations are deductive arguments, it is no easy matter to draw a viable distinction between those arguments that do qualify and those that do not.

Forty years ago a remarkable event occurred. Carl G. Hempel and Paul Oppenheim published an essay, "Studies in the Logic of Explanation," which was truly epoch-making. It set out, with unprecedented precision and clarity, a characterization of one kind of deductive argument that, according to their account, does constitute a legitimate type of scientific explanation. It came later to be known as *the deductive-nomological model*. This 1948 article provided the foundation for the *old consensus* on the nature of scientific explanation that reached its height in the 1960s. A large preponderance of the philosophical work on scientific explanation in the succeeding four decades has occurred as a direct

I should like to express my sincere thanks to Marc Lange for expert bibliographical assistance, and my heartfelt gratitude to Paul Humphreys, Philip Kitcher, and Nicholas Rescher for extremely valuable comments on an earlier draft of this essay. My greatest debt is to Philip Kitcher for his psychological support and intellectual stimulation, without which it would never have been written.

3

or indirect response to this article. If we wish to assess the prospects for a *new consensus* on scientific explanation, this is where we must start. To understand the present situation we need to see how the old consensus came together and how it came apart.

0.1 A Bit of Background

I recall with amusement a personal experience that occurred in the early 1960s. J. J. C. Smart, a distinguished Australian philosopher, visited Indiana University where I was teaching at the time. Somehow we got into a conversation about the major unsolved problems in philosophy of science, and he mentioned the problem of scientific explanation. I was utterly astonished—literally, too astonished for words. At the time I considered that problem essentially solved by the deductive-nomological (D-N) account that had been promulgated by R. B. Braithwaite (1953), Carl G. Hempel (Hempel and Oppenheim 1948), Ernest Nagel (1961), and Karl Popper (1935, 1959), among many others—supplemented, perhaps, by Hempel's then recent account of statistical explanation (Hempel 1962). Although this general view had a few rather vocal critics such as N. R. Hanson (1959) and Michael Scriven (1958, 1959, 1962, 1963) it was widely accepted by scientifically minded philosophers; indeed, it qualified handily as the received view. What is now amusing about the incident is my naïveté in thinking that a major philosophical problem had actually been solved, but my attitude did reflect the then current almost complete consensus.

On one fundamental issue the consensus has remained intact. Philosophers of very diverse persuasions continue to agree that a fundamental aim of science is to provide explanations of natural phenomena. During the last forty years, few (if any) have voiced the opinion that the sole aims of science are to describe, predict, and control nature—that explanation falls into the domains of metaphysics or theology. It has not always been so. Twentieth-century scientific philosophy arose in a philosophical context dominated by post-Kantian and post-Hegelian German idealism. It was heavily infused with transcendental metaphysics and theology. The early logical positivists and logical empiricists saw it as part of their mission to overcome such influences. As philosophers of science they were eager to expunge from science any contamination by super-empirical factors arising out of these philosophies. One such item was teleology, whether in the form of an appeal to the will of a supernatural being who created and continues to direct the course of nature, or in the form of such empirically inaccessible agencies as entelechies and vital forces. In that historical context many metaphysically inclined philosophers argued that there could be no *genuine* explanation of any fact of nature that did not involve an extra-empirical appeal. They thought of explanation anthropomorphically in terms of the sort of 'human understanding' that always appeals to purposes. Many scientific philosophers (as well as philosophical scientists) reacted to this attitude by denying that science is in any way concerned with explanation. Those who did admit that science can offer explanations were eager to make it clear that explanation is nothing more than some special kind of description – it does not demand anything beyond the sphere of empirical knowledge.[1] The classic 1948 Hempel-Oppenheim paper, which will serve as our main point of departure, clearly illustrates this approach.

In recent decades there has been quite general agreement that science can tell us not only *what*, but also *why*. It is possible – in principle and often in practice – to furnish scientific explanations of such facts as the destruction of the space-shuttle *Challenger*, the extinction of the dinosaurs, the coppery color of the moon during total eclipse, and countless other facts, both particular and general. By means of these explanations, science provides us with genuine understanding of the world.

The philosophers who were most instrumental in forging the old consensus – the *logical empiricists* – looked upon the task of philosophy as the construction of *explications* of fundamental concepts. The clearest expression of that goal was given by Rudolf Carnap (1950, 1962, chap. 1; see also Coffa 1973). The concept we are attempting to explicate – in our case, *scientific explanation* – is known as the *explicandum*. This concept, which is frequently used by scientists and by others who talk about science, is vague and, possibly, ambiguous; the job of the philosopher is to provide a clear and exact concept to replace it. The resulting concept is known as the *explicatum*. The process of explication has two stages: first, the explicandum must be clarified sufficiently for us to know what concept it is that we are trying to explicate; second, an exact explicatum must be precisely articulated. Carnap specifies four criteria according to which explications are to be judged:

(1) Similarity to the explicandum. If the explicatum does not match the explicandum to a sufficient degree, it cannot fulfill the function of the concept it is designed to replace. A perfect match cannot, however, be demanded, for the explicandum is unclear and the explicatum should be far more pellucid.

(2) Exactness. Unless the explicatum is precise it does not fulfill the purpose of explication, namely, the replacement of an imprecise concept by a precise one.

(3) Fruitfulness. The new concept should enable us to say significant things and have important insights. One of the main benefits of philosophical analysis should be to deepen our understanding of the nature of science.

(4) Simplicity. The explicatum should be as simple as requirements (1)-(3) permit. Simplicity often accompanies systematic power of concepts. At any rate, simplicity aids in ease of application and avoidance of errors in application.

As Carnap emphatically notes, requirement (1) should not be applied too stringently. The aim is to provide a concept that is useful and clear. In the case of scientific explanation, it is evident that scientists use this concept in a variety of ways, some clear and some confused. Some scientists have claimed, for example, that explanation consists in showing how some unfamiliar phenomenon can be reduced to others that are already familiar; some have equated explanation with something that produces a feeling of intellectual satisfaction. We cannot hope, nor do we want, to capture all of these usages with complete fidelity. The logical empiricists do not indulge in 'ordinary language analysis'—even the ordinary language of scientists—except, perhaps, as a prolegomenon to philosophical analysis.

As already noted, requirement (4) is subservient to its predecessors. Thus, (2) and (3) take precedence: we seek philosophically useful concepts that are formulated with precision. Our discussion of the classic 1948 Hempel-Oppenheim paper in the next section will nicely exemplify the logical empiricist notion of explication. There are, however, several points of clarification that must be made before we turn to consideration of that paper.

First, we must be quite clear that it is *scientific* explanation with which we are concerned. The term "explanation" is used in many ways that have little or nothing to do with scientific explanation (see W. Salmon 1984, 9–11). Scriven once complained that one of Hempel's models of explanation could not even accommodate the case in which one explains with gestures what is wrong with one's car to a Yugoslav garage mechanic who knows no English. Hempel answered, entirely appropriately, that this is like complaining that a precise explication of the term "proof" in mathematics does not capture the meaning of that word as it occurs in such contexts as "86 proof Scotch" and "the proof of the pudding is in the eating" (Hempel 1965, 413). Suitable clarification of the explicandum should serve to forestall objections of that sort.

To seek an explanation for some fact presupposes, of course, that the phenomenon we endeavor to explain did occur—that the putative fact is, indeed, a fact. For example, Immanuel Velikovsky (1950) attempted to 'explain' various miracles reported in the *Old Testament*, such as the sun standing still (i.e., the earth ceasing to rotate) at Joshua's command. Those who are not dogmatically committed to the literal truth of some holy writ will surely require much stronger evidence that the alleged occurrence actually took place before surrendering such basic physical laws as conservation of angular momentum in an attempt to 'explain' it.[2]

To avoid serious confusion we must carefully distinguish between offering an explanation for some fact and providing grounds for believing it to be the case. Such confusion is fostered by the fact that the word "why" frequently occurs in two distinct types of locutions, namely, "Why did X occur?" and "Why should one believe that X occurred?" As an example of the first type, we might ask why Marilyn Monroe died. An answer to this *explanation-seeking* why-question is that she took an overdose of sleeping pills. A full explanation would, of course, identify the particular drug and describe its physiological effects. As an example of the second type, we might ask why we believe that she died. The answer to this *evidence-seeking* why-question, for me at least, is that it was widely reported in the press. Similarly, to take a more scientific example, it is generally believed by cosmologists that the distant galaxies are receding from us at high velocities. The main *evidence* for this hypothesis is the fact that the light from these galaxies is shifted toward the red end of the spectrum, but this red-shift does not *explain* why the galaxies are traveling away from us. The recession of the galaxies is explained on the basis of the "big bang"—the primordial explosion that sent everything flying off in different directions—*not* by the red shift.

It might be supposed that a confusion of evidential facts with explanatory facts is unlikely to arise, but this supposition would be erroneous. In recent years there has been quite a bit of discussion of the so-called *anthropic principle*. According to certain versions of this principle, earlier states of the universe can be explained by the fact that they involved necessary conditions for the later occurrence of life—particularly human life—as we know it. For example, there must have been stars capable of synthesizing nuclei as complex as carbon. It is one thing to infer, from the undisputed fact that human life exists and would be impossible without carbon, that there is some mechanism of carbon synthesis from hydrogen and helium. It is quite another to claim that the existence of human life at present *explains why* carbon was synthesized in stars in our galaxy.[3]

Another fact that sometimes tends to foster the same confusion is the structural similarity of Hempel's well-known *deductive-nomological (D-N)* model of *scientific explanation* (to be discussed in detail in the next section) and the traditional *hypothetico-deductive (H-D)* schema for *scientific confirmation*. It must be kept in mind, however, that the fundamental aims of these two schemas are quite distinct. We *use* well-confirmed scientific hypotheses, laws, or theories to explain various phenomena. The idea behind deductive-nomological explanation is that, given the truth of all of the statements involved—both those that formulate the explanatory facts and the one that asserts the occurrence of the fact-to-be-explained—the logical relation between premises and conclusion shows that the former explain why the latter obtained. The function of the explanation is not to establish (or support) the truth of its conclusion; that is already presupposed when we accept it as a correct explanation. The idea behind the hypothetico-deductive method, in contrast, is that the given logical schema can be employed to provide evidential support for a hypothesis whose truth is being questioned. The statement that is supposed to be supported by hypothetico-deductive reasoning is not the conclusion in the schema, but rather, one of its premises.[4]

Another, closely related, possible source of confusion is the recent popularity of the slogan "inference to the best explanation." As Gilbert Harman has pointed out, we sometimes use the fact that a certain statement, if true, would explain something that has happened as evidence for the truth of that statement (Harman 1965). A detective, attempting to solve a murder, may consider the possible explanations of the crime, and infer that the 'best' one is true. To describe what is going on here it will be useful to appeal to a distinction (made by Hempel and Oppenheim) between potential explanations and actual explanations. A potential explanation has all of the characteristics of a correct—i.e., actual—explanation, except possibly for the truth of the premises. Harman maintains that we canvass the available potential explanations and infer that the 'best' of these is the actual explanation. As in the case of hypothetico-deductive inference, this kind of inference supports the premises of an explanatory argument, not its conclusion, whose truth is taken for granted from the outset. Given the fact that the whole point of the present essay is to discuss a wide variety of views on the nature of scientific explanation, we are hardly in a position at this stage of our investigation to say much of anything about what constitutes 'the best explanation.' And application of this principle of inference obviously presupposes some explication of *explanation*.

0.2 The Received View

Our story begins in 1948 with the publication of the above-mentioned classic article, "Studies in the Logic of Explanation," by Hempel and Oppenheim. This landmark essay provides the initial document of the old consensus concerning the nature of scientific explanation that emerged around the middle of the twentieth century. It is the fountainhead from which the vast bulk of subsequent philosophical work on scientific explanation has flowed—directly or indirectly.

According to that account, a D-N explanation of a particular event is a valid deductive argument whose conclusion states that the event to be explained did occur. This conclusion is known as the *explanandum-statement*. Its premises—known collectively as the *explanans*—must include a statement of at least one general law that is essential to the validity of the argument—that is, if that premise were deleted and *no other change* were made in the argument, it would no longer be valid. The explanation is said to subsume the fact to be explained under these laws; hence, it is often called "the covering law model." An argument fulfilling the foregoing conditions qualifies as a *potential explanation*. If, in addition, the statements constituting the explanans are true, the argument qualifies as a *true explanation* or simply an *explanation* (of the D-N type).

From the beginning, however, Hempel and Oppenheim (1948, 250–51) recognized that not all legitimate scientific explanations are of the D-N variety; some are probabilistic or statistical. In "Deductive-Nomological vs. Statistical Explanation" (1962) Hempel offered his first account of statistical explanation; to the best of my knowledge this is the first attempt by any philosopher to give a systematic characterization of probabilistic or statistical explanation.[5] In "Aspects of Scientific Explanation" (1965) he provided an improved treatment. This account includes two types of statistical explanation. The first of these, the *inductive-statistical (I-S)*, explains particular occurrences by subsuming them under statistical laws, much as D-N explanations subsume particular events under universal laws. There is, however, a crucial difference: D-N explanations subsume the events to be explained deductively, while I-S explanations subsume them inductively. An explanation of either kind can be described as *an argument to the effect that the event to be explained was to be expected by virtue of certain explanatory facts*. In a D-N explanation, the event to be explained is deductively certain, given the explanatory facts (including the laws); in an I-S explanation the event to be explained has high inductive probability relative to the explanatory facts (including the laws).

On Hempel's theory, it is possible to explain not only particular events but also general regularities. Within the D-N model, universal generalizations are explained by deduction from more comprehensive universal generalizations. In the second type of statistical explanation, the *deductive-statistical (D-S)*, statistical regularities are explained by deduction from more comprehensive statistical laws. This type of statistical explanation is best regarded as a subclass of D-N explanation.

Table 1 shows the four categories of scientific explanations recognized by Hempel in "Aspects." However, in their explication of D-N explanation in 1948, Hempel and Oppenheim restrict their attention to explanations of particular facts, and do not attempt to provide any explication of explanations of general regularities. The reason for this restriction is given in the notorious footnote 33:

> The precise rational reconstruction of explanation as applied to general regularities presents peculiar problems for which we can offer no solution at present. The core of the difficulty can be indicated by reference to an example: Kepler's laws, K, may be conjoined with Boyle's law, B, to [form] a stronger law $K.B$; but derivation of K from the latter would not be considered an explanation of the regularities stated in Kepler's laws; rather, it would be viewed as representing, in effect, a pointless "explanation" of Kepler's laws by themselves. The derivation of Kepler's laws from Newton's laws of motion and

Explananda / Laws	Particular Facts	General Regularities
Universal Laws	D-N Deductive-Nomological	D-N Deductive-Nomological
Statistical Laws	I-S Inductive-Statistical	D-S Deductive-Statistical

Table 1

gravitation, on the other hand, would be recognized as a genuine explanation in terms of more comprehensive regularities, or so-called higher-level laws. The problem therefore arises of setting up clear-cut criteria for the distinction of levels of explanation or for a comparison of generalized sentences as to their comprehensiveness. The establishment of adequate criteria for this purpose is as yet an open problem. (Hempel and Oppenheim 1948, 273; future citations, H-O 1948)

This problem is not resolved in any of Hempel's subsequent writings, including "Aspects of Scientific Explanation."

Chapter XI of Braithwaite's *Scientific Explanation* is entitled "Explanation of Scientific Laws," but it, too, fails to address the problem stated in the Hempel-Oppenheim footnote. Indeed, on the second page of that chapter Braithwaite says,

> To explain a law is to exhibit an established set of hypotheses from which the law follows. It is not necessary for these higher-level hypotheses to be established independently of the law which they explain; all that is required for them to provide an explanation is that they should be regarded as established and that the law should follow logically from them. It is scarcely too much to say that this is the whole truth about the explanation of scientific laws . . . (Braithwaite 1953, 343)

It would appear that Braithwaite is prepared to say that the deduction of Kepler's laws from the conjunction of Kepler's laws and Boyle's law — or the conjunction of Kepler's laws and the law of diminishing marginal utility of money (if you accept the latter as an established law) — is a bona fide explanation of Kepler's laws. However, inasmuch as Braithwaite's book does not contain any citation of the Hempel-Oppenheim paper, it may be that he was simply unaware of the difficulty, at least in this precise form. This problem was addressed by Michael Friedman (1974); we shall discuss his seminal article in §3.5 below. It was also treated by John Watkins (1984); his approach will be discussed in §4.10. Since the same problem obviously applies to D-S explanations, it affects both sectors in the right-hand column of Table 1.

The 1948 Hempel-Oppenheim article marks the division between the prehistory and the history of modern discussions of scientific explanation.[6] Hempel's 1965 "Aspects" article is the central document in the hegemony (with respect to scientific explanation) of logical empiricism, which held sway during roughly the third quarter of the present century. Indeed, I shall use the phrase *the received view* to refer to accounts similar to that given by Hempel in "Aspects." According to the received view, I take it, every legitimate scientific explanation belongs to one of the four sectors of Table 1. As we have seen, the claim of the received view to a comprehensive theory of scientific explanation carries a large promissory note regarding explanations of laws.

The First Decade (1948–57)
Peace in the Valley
(but Some Trouble in the Foothills)

With hindsight we can appreciate the epoch-making significance of the 1948 Hempel-Oppenheim paper; as we analyze it in detail we shall see the basis of its fertility. Nevertheless, during the first decade after its appearance it had rather little influence on philosophical discussions of explanation. To the best of my knowledge only one major critical article appeared, and it came at the very end of the decade (Scheffler 1957); it was more a harbinger of the second decade than a representative of the first. Indeed, during this period not a great deal was published on the nature of scientific explanation in general (in contrast to explanation in particular disciplines).

Braithwaite's (1953) might come to mind as a possible major exception, but we should not be misled by the title. In fact this book contains hardly any explicit discussion of the topic. Braithwaite remarks at the outset that "to understand the way in which a science works, and the way in which it provides explanations of the facts which it investigates, it is necessary to understand the nature of scientific laws, and what it is to establish them" (1953, 2). He then proceeds to discuss at length the nature of scientific deductive systems, including those that involve statistical laws, as well as those that involve only universal laws. Throughout this detailed and illuminating discussion he seems to be assuming implicitly that scientific explanation consists in somehow embedding that which is to be explained in such a deductive system. In adopting this view he appears to be anticipating the Friedman-Kitcher global unification approach, which will be discussed in §3.5 below. However, he has little to say explicitly about the relationship between deductive systems and scientific explanation.[1]

The final two chapters take up some specific issues regarding scientific explanation, and in the course of these chapters Braithwaite makes a few general remarks in passing. For example, the penultimate chapter opens with the statement, "Any proper answer to a 'Why?' question may be said to be an explanation of a sort" (319). In the final chapter he remarks, similarly, "an explanation, as I understand the use of the word, is an answer to a 'Why?' question which gives some intellectual satisfaction" (348–49). These comments are, without doubt, intended for construal in terms of his foregoing discussion of formal systems, but he does not spell out the connections. In the absence of explicit analyses of the nature of why questions, of what constitutes a "proper answer," or of the notion of "intellectual satisfaction," such passing remarks, however suggestive, leave much to be desired. The fact that Braithwaite's book nowhere cites the Hempel-Oppenheim article is eloquent testimony to the neglect of that essay during the first decade.

During this decade interest focused chiefly on two sets of special issues that had been sparked by earlier work. One of these concerned the nature of historical explanation, and the question of whether historical explanations must involve, at least implicitly, appeals to general laws. Much of this discussion took as its point of departure an earlier paper by Hempel (1942). The other dealt with the question of teleological or functional explanation; it came out of the longstanding controversy over mechanism vs. teleology (see H-O 1948, §4). On this specific issue, as we shall see, Braithwaite's book does provide significant contributions. We shall return to these special topics in §1.2 and §1.3, respectively, and we shall find an important connection between them.

1.1 The Fountainhead: The Deductive-Nomological Model

The 1948 Hempel-Oppenheim paper makes no pretense of explicating anything other than D-N explanations of particular occurrences — represented by the upper left-hand sector of Table 1. It will be useful to look in some detail at their treatment of this case. We must distinguish, in the first place, between the general conditions of adequacy for any account of this type of explanation, as laid down in Part I, and the actual explication spelled out in Part III.

The general conditions of adequacy are divided into two groups, logical and empirical. Among the logical conditions we find

    (1) the explanation must be a valid deductive argument,
    (2) the explanans must contain essentially at least one general law,
    (3) the explanans must have empirical content.

The only empirical condition is:

    (4) the sentences constituting the explanans must be true.

Although these criteria may seem simple and straightforward, they have been called into serious question. We shall return to this matter a little later.

The general notion of D-N explanation can be represented in the following schema offered by Hempel and Oppenheim, where the arrow signifies deductive entailment. It should also be noted that these criteria of adequacy are meant to apply to D-N explanations of general regularities even though Hempel and Oppenheim do not attempt to provide an explicit explication of explanations of this type. Since the derivation of a narrower generalization (e.g., the behavior of double stars) from a more comprehensive theory (e.g., celestial mechanics) does not

$C_1, C_2, \ldots, C_k$ Statements of antecedent
conditions

                                          Explanans

$L_1, L_2, \ldots, L_r$ General laws

      E   Description of the empirical
phenomenon to be explained   Explanandum

require any antecedent conditions, they deliberately refrain from requiring that the explanans contain any statements of antecedent conditions.

One of the most vexing problems arising in this context is the characterization of law-sentences. It obviously has crucial importance for the D-N model, as well as for any covering law conception of scientific explanation. Following a strategy introduced by Nelson Goodman (1947), Hempel and Oppenheim (1948, 264–70) attempt to define the broader notion of a *lawlike sentence*. Only true sentences are classified as law-sentences; lawlike sentences have all the characteristics of law-sentences, with the possible exception of truth. Thus every law-sentence is a lawlike sentence, but not all lawlike sentences are laws. Informally, lawlike sentences have four properties:

(1) they have universal form,
(2) their scope is unlimited,
(3) they do not contain designations of particular objects, and
(4) they contain only purely qualitative predicates.

Let us consider the reasons for requiring these characteristics. With regard to (1) and (2) it is intuitively plausible to expect laws of nature to be general laws whose variables range over the entire universe. Newton's laws of universal gravitation and motion apply to all bodies in the universe, and their scope is not restricted in any way. These are paradigms of lawlike statements. However, an apparently universal statement, such as "All Apache pottery is made by women," would not qualify as lawlike because its scope is restricted. Likewise, the statement, "All living things contain water," if tacitly construed to be restricted to living things on earth, would not qualify as lawlike. In contrast, however, "All pure gold is malleable"—though it may appear to have a scope limited to golden objects—is nevertheless a universal generalization of unlimited scope, for it says of each object in the universe that, if it consists of gold, it is malleable. The distinction among the foregoing examples between those that qualify as lawlike and those that do not relates to characteristic (3). The statement about Apache pottery makes explicit reference to a particular group of people, the Apache. The statement about living things, if construed as suggested, refers implicitly to our particular planet.[2]

Why does it matter, with respect to lawlikeness, whether a statement refers to a particular of some sort—a particular time, place, object, person, group, or nation? Consider a simple example. Suppose it happens to be true (because I like golden delicious apples) that all of the apples in my refrigerator are yellow. This statement involves reference to a particular person (me), a particular thing (my refrigerator), and a particular time (now). Even given my taste in apples it is not impossible for my refrigerator to contain apples of different colors. Moreover, there is no presumption that a red delicious apple would turn yellow if it were placed in my refrigerator.

The problem that arises in this context is to distinguish between laws and *accidental generalizations*. This is a crucial issue, for laws have explanatory force, while accidental generalizations, *even if they are true*, do not. It obviously is no explanation of the color of an apple that it happens to reside in my refrigerator at some particular time.

If a statement is to express a law of nature it must be true. The question is, what characteristics, in addition to truth, must it possess? *Generality* is one such characteristic: laws must apply universally and they must not contain special provisions or exceptions for particular individuals or groups. The *ability to support counterfactuals* is another: they must tell us what would happen if. . . . If this table salt were placed in water, it would dissolve. If this switch were closed, a current would flow in this circuit.[3] *Modal import* is another: laws delineate what is necessary, possible, or impossible. We are not talking about logical modalities, of course; we are concerned with what is *physically* necessary, possible, or impossible. According to relativity theory it is physically impossible to send a signal faster than light *in vacuo*; according to the first law of thermodynamics it is physically impossible to construct a perpetual motion machine (of the first type). Accidental generalizations, even if true, do not support counterfactuals or possess modal import.

Even if a given statement does not contain explicit designations of particular objects, it may involve implicit reference to one or more particulars. Such references may be hidden in the predicates we use. Terms like "lunar," "solar," "precolumbian," and "arctic," are obvious examples. Because such terms refer to particulars they do not qualify as purely qualitative. By stipulating, in property (4) above, that laws contain only purely qualitative predicates, this sort of implicit reference to particulars, is excluded. Properties (3) and (4) are designed to rule out as accidental those universal generalizations that contain either explicit or implicit reference to particulars.

As Hempel and Oppenheim are fully aware, the prohibition against reference to particulars they impose is extremely stringent. Under that restriction, neither Galileo's law of falling bodies (which refers explicitly to the earth) nor Kepler's laws of planetary motion (which refer explicitly to our solar system) would qualify as laws or lawlike statements. As we shall see, because of this consideration they distinguish between fundamental and derived laws. The foregoing restrictions apply only to the fundamental laws. Any universal statement that can be deduced from fundamental laws qualifies as a derived law.

Yet, in spite of their careful attention to the problem of distinguishing between lawful and accidental generalizations, Hempel and Oppenheim did not succeed in explicating that distinction. Consider the following two statements:

(i) No signal travels faster than light.

(ii) No gold sphere has a mass greater than 100,000 kg.

Let us suppose, for the sake of argument, that both are true. Then we have two true (negative) universal generalizations. Both have universal form. Neither is restricted in scope; they refer, respectively, to signals and gold spheres anywhere in the universe at any time in its history—past, present, or future. Neither makes explicit reference to any particulars. Both statements satisfy characteristics (1)-(3). One might argue that the predicate "having mass greater than 100,000 kg" is not purely qualitative, since it contains a reference to a particular object—namely, the international prototype kilogram. But this difficulty can be avoided by expressing the mass in terms of atomic mass units (which refer, not to any particular object, but to carbon-12 atoms in general). Thus, with (ii) suitably reformulated, we have two statements that satisfy characteristics (1)-(4), one of which seems patently lawful, the other of which seems patently accidental. The contrast can be heightened by considering

(iii) No enriched uranium sphere has a mass greater than 100,000 kg.

Since the critical mass for enriched uranium is just a few kilograms, (iii) must be considered lawful.

Both statements (i) and (iii) have modal import, whereas (ii) does not. It is physically impossible to send a message faster than light and it is physically impossible to fabricate an enriched uranium sphere of mass greater than 100,000 kg. It is not physically impossible to fabricate a gold sphere of mass greater than 100,000 kg.[4] Likewise, statements (i) and (iii) support counterfactuals, whereas (ii) does not. If something were to travel faster than light it would not transmit information.[5] If something were a sphere with mass greater than 100,000 kg it would not be composed of enriched uranium. In contrast, we cannot legitimately conclude from the truth of (ii) that if something were a sphere with mass greater than 100,000 kg, it would not be composed of gold. We cannot conclude that if two golden hemispheres with masses greater than 50,000 kg each were brought together, they would explode, suffer gravitational collapse, undergo severe distortion of shape, or whatever, instead of forming a sphere.

Lawfulness, modal import, and support of counterfactuals seem to have a common extension; statements either possess all three or lack all three. But it is extraordinarily difficult to find criteria to separate those statements that do from those that do not. The three characteristics form a tight little circle. If we knew which statements are lawful, we could determine which statements have modal import and support counterfactuals. But the way to determine whether a statement has modal import is to determine whether it is a law. The same consideration applies to support of counterfactuals; to determine which statements support counterfactuals we need to ascertain which are laws.[6] The circle seems unbroken. To determine to which statements any one of these characteristics applies we need to be able to determine to which statements another of them applies.

There are, of course, a number of differences between statements (i) and (iii) on the one hand and statement (ii) on the other. For example, I am much less confident of the truth of (ii) than I am of (i) or (iii). But this is a psychological statement about my state of belief. However, we are assuming the truth of all three statements. Given that all three are true, is there any objective difference in their status, or is the sole difference psychological? Again, (i) and (iii) fit closely with a well-integrated body of physical theory, while (ii) does not.[7] But given that all three are true, is this more than an epistemic difference?[8] Further, there are differences in the ways I might come to know the truth of (ii), as opposed to coming to know the truth of (i) and (iii). But is this more than an epistemic or psychological difference? Still further, I would much more readily give up my belief in (ii) than I would my belief in (i) or (iii).[9] But is this more than a pragmatic difference? The unresolved question is this: is there any objective distinction between laws and true accidental generalizations?[10] Or is the distinction wholly psychological, epistemic, or pragmatic?

In his 1953 book, Braithwaite places considerable emphasis upon the nature of laws and their place in science. He writes, "In common with most of the scientists who have written on philosophy of science from Ernst Mach and Karl Pearson to Harold Jeffreys, I agree with the principal part of Hume's thesis—the part asserting that universals of law are objectively just universals of fact, and that in nature there is no extra element of necessary connexion" (1953, 294). In chapter IX he defends the view that "the difference between universals of law and universals of fact [lies] in the different roles they play in our thinking rather than in any difference in their objective content" (294–95).[11]

The most ambitious attempt by any of the logical empiricists to deal with these problems concerning the nature of laws was given by Hans Reichenbach (1954),[12] the year just after the publication of Braithwaite's book. It had been anticipated by his discussion of the same topics in his symbolic logic book (1947, chap. VIII). Reichenbach's very first requirement on law-statements makes the distinction between laws and accidental generalizations an epistemic one, for it refers explicitly to the types of evidence by which such statements are supported. It should be remarked, incidentally, that Reichenbach was not addressing these problems in the context of theories of scientific explanation.

The problem of characterizing law-statements is one that has not gone away. Skipping ahead to subsequent decades, we may note that Ernest Nagel's magnum opus on scientific explanation, published near the beginning of the second decade, has a sensitive and detailed discussion of this problem, but one that remains inconclusive.[13] Around the beginning of the third decade, Nicholas Rescher's book *Scientific Explanation* offers an extended discussion which concludes that lawfulness does not reflect objective factors in the world, but rather rests upon our imputations, and is consequently mind-dependent (1970, 97-121; see also Rescher 1969). In the fourth decade, to mention just one example among many, Brian Skyrms (1980) offers a pragmatic analysis. The fifth decade will see the publication of an extremely important work on the subject, *Laws and Symmetries*, by Bas van Fraassen. But let us return to the first decade.

To carry out their precise explication, Hempel and Oppenheim introduce a formal language in which scientific explanations are supposed to be formulated. It is a standard first order functional calculus without identity, but no open sentences are allowed. All individual variables are quantified, so generality is always expressed by means of quantifiers. Two semantic conditions are imposed on the interpretation of this language: First, the range of the individual variables consists of all physical objects in the universe or of all spatio-temporal locations; this ensures that requirement (2) on lawlike statements — that their scope be unlimited — will be fulfilled, for there is no limit on the range of the variables that are universally (or existentially) quantified. Second, the primitive predicates are all purely qualitative; this feature of the interpretation of the language is, of course, a direct reflection of the fourth requirement on lawlike statements. The explication of D-N explanation of particular occurrences is given wholly in semantic terms.

Before going into the details of the formal language, we must acknowledge a fundamental problem regarding the second of the foregoing semantic conditions, namely, the concept of a purely qualitative predicate. In his well-known book *Fact, Fiction, and Forecast* (1955), Nelson Goodman poses what he calls "the new riddle of induction" in terms of two predicates, "grue" and "bleen," that he constructs for that purpose. Select quite arbitrarily some future time *t* (say the beginning of the twenty-first century). "The predicate 'grue' applies to all things examined before *t* just in case they are green but to other things just in case they are blue" (1955, 74). "Bleen" applies to things examined before *t* just in case they are blue but to other things just in case they are green (1955, 79). The question Goodman poses is whether we should inductively project that twenty-first century emeralds will be green or that they will be grue. The same problem had originally been posed by Goodman in 1947.

In an answer to Goodman's query, Carnap maintained that "grue" and "bleen," in contrast to "blue" and "green," are not purely qualitative predicates, because of the reference to a particular time in their definitions. He proposes to resolve Goodman's problem by restricting the predicates of his languages for confirmation theory to purely qualitative ones (Carnap 1947).[14] Goodman demurs:

. . . the argument that the former but not the latter are purely qualitative seems to me quite unsound. True enough, if we start with "blue" and "green," then "grue" and "bleen" will be explained in terms of "blue" and "green" and a temporal term. But equally truly, if we start with "grue" and "bleen," then "blue" and "green" will be explained in terms of "grue" and "bleen" and a temporal term; "green," for example, applies to emeralds examined before time *t* just in case they are grue, and to other emeralds just in case they are bleen. Thus qualitativeness is an entirely relative matter.(1947)

It is now generally conceded that Carnap's attempt to characterize purely qualitative predicates was inadequate to deal with the problem Goodman raised. Many philosophers (including this one – (W. Salmon 1963)) have tried to make good on the distinction Carnap obviously had in mind. Whether any of these other efforts have been successful is a matter of some controversy; at any rate, no particular solution has gained general acceptance.

Our discussion, so far, has been largely preparatory with respect to the official Hempel-Oppenheim explication. We may now return to the formal language offered by Hempel and Oppenheim. Several different types of sentences must be distinguished. To begin, an *atomic sentence* is one that contains no quantifiers, no variables, and no sentential connectives. It is a sentence that attributes a particular property to a given individual (e.g., "George is tall") or asserts that a particular relation holds among two or more given individuals (e.g., "John loves Mary"). A *basic sentence* is either an atomic sentence or the negation of an atomic sentence; a basic sentence contains no quantifiers, no variables, and no binary sentential connectives. *Singular* (or *molecular*) *sentences* contain no quantifiers or variables, but they may contain binary sentential connectives (e.g., "Mary loves John or Mary loves Peter"). A *generalized sentence* contains one or more quantifiers followed by an expression containing no quantifiers (e.g., "All humans are mortal"). Since any sentence in first order logic can be transformed into *prenex normal form*, any sentence containing quantifiers can be written as a generalized sentence. *Universal sentences* are generalized sentences containing only universal quantifiers. A generalized (universal) sentence is *purely generalized (universal) if it contains no proper names of individuals. A generalized (universal)* sentence is an *essentially generalized (universal)* sentence that is not equivalent to any singular sentence. With these definitions in hand we can proceed to explicate the fundamental concepts involved in scientific explanation.

The first concept with which we must come to terms is that of a law of nature, and, as we have seen, it is one of the most problematic. Hempel and Oppenheim distinguish between *lawlike sentences* and genuine laws, and also between *fundamental* and *derivative* laws. The following series of definitions is offered:[15]

(7.3a) A *fundamental lawlike sentence* is any purely universal sentence; a *fundamental law* is purely universal and true.

(7.3b) A *derivative law* is a sentence that is essentially, but not purely, universal and is deducible from some set of fundamental laws.

(7.3c) A *law* is any sentence that is either a fundamental or a derived law.

We have already canvassed the fundamental problems encountered in this characterization of laws.

Interestingly, the concept of law does not enter into the formal explication of D-N explanation; instead, the notion of *theory* is employed.

(7.4a) A *fundamental theory* is any sentence that is purely generalized and true.

(7.4b) A *derivative theory* is any sentence that is essentially, but not purely, generalized and is derivable from fundamental theories.

(7.4c) A *theory* is any fundamental or derivative theory.

Note that the concept of a theory-like sentence is not introduced.

According to the foregoing definitions, every law is a theory and every theory is true. As the term "theory" is used in this context, there is no presumption that theories refer to unobservable entities, or that they involve any sort of special theoretical vocabulary. The difference between laws and theories is simply that theories may contain existential quantifiers, while laws contain only universal quantifiers. Clearly, many of the scientific laws or theories that are employed in explanation contain existential quantifiers. To say, for example, that every comet has a tail, that every atom has a nucleus, or that every mammal has a heart, involves a universal quantifier followed by an existential quantifier—i.e., for every x there is a y such that . . . Hempel and Oppenheim say nothing about the order in which quantifiers must occur in theories. That leaves open the interesting question of whether explanatory theories may have existential quantifiers preceding all of the universal quantifiers, or whether explanatory theories need contain any universal quantifiers at all.[16] It is perhaps worth explicit mention in this context that universality and generality are not coextensive. Existentially quantified statements are general in the sense that they involve variables having the universe as their range. To say, "there exists an x such that . . . " means that within the whole domain over which x ranges there is at least one object such that. . . . Such statements have generality without being universal. The question remains whether universality is a necessary requirement for explanatory theories, or whether generality is sufficient. As we shall see in connection with the next set of formal definitions, Hempel and Oppenheim are willing to settle for the latter alternative.

We have finally arrived at the stage at which Hempel and Oppenheim offer their formal explication of scientific explanation. The concept of a potential explanation comes first:

(7.5) $<T,C>$ is a potential explanans of E (a singular sentence) *only if*
   (1) T is essentially general and C is singular, and
   (2) E is derivable from T and C jointly, but not from C alone.

It would be natural to suppose that (7.5) would constitute a definition of "potential explanans," but Hempel and Oppenheim are careful to point out that it provides only a necessary condition. If it were taken as sufficient as well, it would leave open the possibility that "any given particular fact could be explained by means of any true lawlike sentence whatever" (H-O 1948, 276). They offer the following example. Let the explanandum-statement E be "Mount Everest is snowcapped" and let the theory T be "All metals are good conductors of heat." Take a singular sentence T_s that is an instance of $T-$e.g., "If the Eiffel Tower is metal it is a good conductor of heat." Now take as the singular sentence C the sentence T_s implies $E-$i.e., "If the fact that the Eiffel Tower is made of metal implies that it is a good conductor of heat, then Mount Everest is snowcapped." Because E is true, C must be true, for C is a material conditional statement with a true consequent. Thus,

$$\vdash C = T_s \supset E \qquad \text{definition}$$
$$T_s \supset E \qquad \text{assumption}$$
$$\vdash T \supset T_s \qquad \text{instantiation}$$
$$T \supset E \qquad \text{hypothetical syllogism}$$
$$\vdash C \supset (T \supset E) \qquad \text{conditional proof}$$
$$\vdash C . T \supset E \qquad \text{tautology}$$

It is evident that C does not, by itself, entail E. Therefore $<T,C>$ satisfies (7.5). But it is manifestly absurd to claim that the law about metals being good conductors of heat is the key law in the explanation of snow on Mount Everest.

The obvious difficulty with this example is that C's truth can be fully certified only on the basis of the truth of E. Evidently, some restriction must be placed on the singular sentence C that is to serve as the statement of antecedent conditions in the explanans. If knowing that the explanandum-statement is true is the only way to establish the truth of C, then in some important sense, in appealing to C, we are simply using E to explain E. Indeed, given that T is true, there must be some way to establish the truth of C without appealing to E. Hempel and Oppenheim formulate the needed restriction as follows:

   (3) T must be compatible with at least one class of basic sentences which has C but not E as a consequence.

That is to say, given that the theory T is true, there must be some way to verify that C is true without also automatically verifying E as well. Adding (3) to the necessary conditions stated in (7.5) gives

(7.8) $<T,C>$ is a potential explanans of E (a singular sentence) *iff*
   (1) T is essentially general and C is singular, and

(2) E is derivable from T and C jointly, but not from C alone.

(3) T must be compatible with at least one class of basic sentences which has C but not E as a consequence.[17]

With this definition of "potential explanans" it is a small step to the official explication of "explanans," and hence, "explanation."

(7.6) $<T,C>$ is an explanans of E (a singular sentence) *iff*

    (1) $<T,C>$ is a potential explanans of E

    (2) T is a theory and C is true.

Taken together, the explanans $<T,C>$ and the explanandum E constitute an explanation of E. This completes the Hempel-Oppenheim explication of *D-N explanation of a particular fact.*

Given the great care with which the foregoing explication was constructed, it would be easy to surmise that it is technically correct. Jumping ahead to the next decade for a moment, we find that such a supposition would be false. As Rolf Eberle, David Kaplan, and Richard Montague (1961) showed (roughly), on the foregoing explication any theory T can explain any fact E, where T and E have no predicates in common, and are therefore, intuitively speaking, utterly irrelevant to one another. Suppose, for example, that T is "(x)Fx" (e.g., "Everyone is imperfect.") and E is "Ha" (e.g., "C. G. Hempel is male.").[18] We can formulate another theory T' that is a logical consequence of T:

$$T' =_{df} (x)(y)[Fx \lor (Gy \supset Hy)]$$

T' is of purely universal form, and, on the assumption that T is true, it is true as well. As a singular sentence, take

$$C =_{df} (Fb \lor \sim Ga) \supset Ha$$

For the sake of our concrete interpretation, we can let "Gx" mean "x is a philosopher" and let "b" stand for W. V. Quine. It can now be shown that $<T',C>$ constitutes an explanans of E.

(1)	$(x)(y)[Fx \lor (Gy \supset Hy)]$	premise (T')
(2)	$(Fb \lor \sim Ga) \supset Ha$	premise (C)
(3)	$(Fb \supset Ha) . (\sim Ga \supset Ha)$	equivalent to (2)
(4)	$Fb \supset Ha$	simplification (3)
(5)	$\sim Ga \supset Ha$	simplification (3)
(6)	$Fb \lor (Ga \supset Ha)$	instantiation (1)
(7)	$\sim Fb \supset (Ga \supset Ha)$	equivalent to (6)
(8)	$Fb \lor \sim Fb$	tautology
(9)	$Ha \lor (Ga \supset Ha)$	dilemma (4, 7, 8)
(10)	$Ha \lor \sim Ga \lor Ha$	equivalent to (9)
(11)	$\sim Ga \lor Ha$	equivalent to (10)

(12)	Ga ⊃ Ha	equivalent to (11)
(13)	Ga v ~ Ga	tautology
(14)	Ha	dilemma (5, 12,13)

As we have seen, T′ is essentially general, C is singular, and E is derivable from T′ and C. Hence, conditions (1) and (2) of (7.8) are satisfied. Now, consider the set of basic sentences {~ Fb, Ga}; obviously it does not entail E (i.e., Ha). But it does entail C, as follows:

(1)	~ Fb . Ga	premise
(2)	(~ Fb . Ga) v Ha	addition (1)
(3)	~ (~ Fb . Ga) ⊃ Ha	equivalent to (2)
(4)	(Fb v ~ Ga) ⊃ Ha	DeMorgan (3)

Thus, condition (3) of (7.8) is also fulfilled; < T′,C > is an explanans for E. We should remember that < T,C > has not been shown to be an explanans for E, so it has not been shown that any arbitrarily chosen theory explains any particular fact (as long as they share no predicates). But what has been shown is something like this: from the law of diminishing marginal utility of money (which, for the sake of argument, I take to be true), we can deduce a theory which, together with certain singular statements, provides an explanation of the explosion of the *Challenger* space-shuttle vehicle. It can hardly be doubted that Eberle, Kaplan, and Montague offered a recipe (in fact, several recipes) for constructing counterexamples that are damaging to the Hempel-Oppenheim explication.

To assess the nature of the damage, let us go back to Hempel and Oppenheim's set of necessary conditions (7.5), which at first blush looks like a suitable explication of *potential D-N explanans*. Had they offered it as such, their explication would have been technically defective, for it would have been vulnerable to the Mount Everest counterexample. However, it would not have posed a profound problem for their enterprise, since the technical defect could have been repaired by technical tinkering, namely, by adding the third condition of (7.8). Fortunately, they anticipated the problem and found a solution before their essay was published.[19]

The same attitude should, I believe, be taken to the problem discovered 13 years later by Eberle, Kaplan, and Montague. They pointed out a genuine technical defect, but again, it was one that could be repaired by technical tinkering. Indeed, in the same volume of the same journal, Kaplan (1961) provided one way of making the repair. Shortly thereafter another was offered by Jaegwon Kim (1963).

It is worth noting, I believe, that both the Mount Everest example and the Eberle-Kaplan-Montague type examples exploit a well-known feature of standard truth-functional logic. This feature is the principle of addition that allows one to deduce "p v q" from "p" — where "q" can be any arbitrary sentence whatever. This

principle is closely related to one of the so-called paradoxes of material implication, namely, the fact that if "p" is true then it is materially implied by any arbitrary sentence whatever—e.g., "~q"—so that, given "p" we can infer
"~q ⊃ p" (which is equivalent to "p v q"). In the Mount Everest example, we
chose C as $T_s \supset E$, which we were prepared to assert on the basis of the truth
of E. In the Eberle-Kaplan-Montague example we chose C as "(Fb v
~Ga) ⊃ Ha"—where "Ha" is E. In addition, from the theory "(x)Fx" we derived
"(x)(y)[Fx v (Gy ⊃ Hy)]" which is equivalent to "(x)Fx v (y)(Gy ⊃ Hy)."
Clearly the source of difficulty is that these moves are precisely the sorts that allow the introduction of irrelevancies—they are the very principles that are excluded in relevance logics. The technical problem with the Hempel-Oppenheim
explication is simply to find ways of blocking them in the context of explanation.
Kaplan and Kim have shown how to do that. Kim's revision consists in adding
a further requirement to those already included in (7.8):

(4) E must not entail any conjunct of the conjunctive normal form of C.[20]

In 1965 Hempel could be quite sanguine about the technical details of the corrected explication (Hempel 1965, 294–95).

It should be emphatically noted that the official explication—as given in the
augmented definition (7.8) is very different from the set of requirements of adequacy given in the first section of the Hempel and Oppenheim essay. This is to
be expected. The informal conditions of adequacy are part of the clarification of
the explicandum; the explication is the formal definition of the improved concept
that is to replace the original vague concept of explanation. What is perhaps surprising is the complexity of the formal explication.

Now that we have looked at the nitty-gritty details of the Hempel-Oppenheim
explication, let us return to a consideration of the fundamental philosophical issues to which it gives rise. Where does this discussion leave us? First, there is
a two-part agenda, explicitly stated in that essay, of items that are needed to complete the received view:

(1) Explications of one or more models of probabilistic or statistical explanation.

(2) An explication of D-N explanation of laws.

Second, there are, as we have seen, two serious lacunae in the Hempel-
Oppenheim explication of D-N explanation of particular facts:

(3) A satisfactory explication of the concept of a purely qualitative predicate.

(4) A satisfactory explication of the concept of a law of nature.

Third, as we shall see in greater detail when we discuss the second decade,
the Hempel-Oppenheim essay advances several important philosophical theses
that have been the subject of much controversy:

(5) The inferential conception of scientific explanation—i.e., the thesis that all legitimate scientific explanations are arguments of one sort or another. This thesis is involved in Hempel and Oppenheim's first condition of adequacy for scientific explanations. A number of subsequent authors including Michael Scriven, Richard Jeffrey, Bas van Fraassen, Peter Railton, and I—have rejected this inferential conception.

(6) The covering law conception of scientific explanation[21] —i.e., the view that every fully articulated adequate scientific explanation contains one or more laws in an essential fashion. This thesis is set forth in Hempel and Oppenheim's second condition of adequacy. As we shall see, this view is also rejected by a number of authors—including Scriven and van Fraassen. Thomas S. Kuhn strenuously denies it.[22] So do many others who are primarily concerned with the nature of explanation in history.

(7) The explanation/prediction symmetry thesis. According to the strong form of this thesis, as set forth in the Hempel-Oppenheim article, any correct D-N explanation could serve, in appropriate circumstances, as a scientific prediction; conversely, any deductive scientific prediction could, in appropriate circumstances, serve as a D-N explanation. This thesis, which Hempel later extended to I-S explanation as well, has been widely disputed. Hanson, Israel Scheffler, Scriven, van Fraassen, and I—among many others—have been unwilling to accept it.

(8) The role of causality in scientific explanation. Hempel and Oppenheim casually identify causal explanation with D-N explanation (H-O 1948, 250), but their official explication makes no reference to any causal requirements. In "Aspects," Hempel explicitly rejects the idea that causality plays any essential explanatory role (1965, 352). The question is whether to follow Hempel in "Aspects" in expunging causality from explanation altogether, or to find a place for causality, as Scriven, Railton, James H. Fetzer, Paul Humphreys, and I—among many others—have urged. As we shall see, this issue has been the subject of considerable subsequent discussion.

(9) Literal truth of the explanans. This is Hempel and Oppenheim's fourth condition of adequacy. Many authors have maintained that this requirement is too strong—that it should be replaced by a requirement of high degree of confirmation, or by a requirement of approximate truth. For various reasons this requirement has been rejected by Nancy Cartwright, Kuhn, and Larry Laudan.

(10) The possibility of quasi-formal models of explanation. The early logical empiricists seemed confident that many fundamental methodological concepts—such as confirmation and explanation—are amenable to formal semantical explication, more or less as attempted in the Hempel-Oppenheim paper. It now seems to many—including Peter Achinstein and me—that explications of that sort are bound to turn out to be inadequate. Even Hempel in his later writings on scientific

explanation eschews the degree of formality employed in the Hempel-Oppenheim explication.

Our review of the essay by Hempel and Oppenheim leaves no doubt whatever about its richness as a source of material for philosophical discussions of the nature of scientific explanation in the second half of the twentieth century. Can anyone seriously dispute the propriety of its designation as the fountainhead?

1.2 Explanation in History and Prehistory

In the decade following the publication of the Hempel-Oppenheim paper, as I pointed out above, not a great deal appeared in print on general problems concerning the nature of scientific explanation.[23] During this time, however, quite a good deal was written on the nature of explanation in history, stimulated, to a large degree, by Hempel's essay, "The Function of General Laws in History" (1942). The focus of much of the discussion was the question of whether historical explanations can or should embody general laws. This literature is directly pertinent to our discussion only if history is considered a science. Since I am not firmly committed to any particular view on this matter, and know very little about history, I shall not pursue the issue here. I refer the reader to Hempel's 1942 article, to §7 of "Aspects," and to the excellent bibliography of that essay.

Whether history is classified as a science or not, there can be no doubt that some sciences have essential historical aspects. Cosmology, geology, and evolutionary biology come immediately to mind. Archaeology (often called "prehistory") is somewhat ambiguous, for it has been pursued, not infrequently, as a humanistic, rather than scientific, discipline.

During the 1950s, an important development occurred in archaeology, namely, the origin of an influential movement called the *New Archaeology*. The main thrust of this movement is the attempt to make archaeology a bona fide science. When Lewis Binford, one of its major founders, was a student at the University of Michigan, Leslie White—a distinguished cultural anthropologist and a teacher of Binford—told him that, to find out what it means to be scientific, he should read some philosophy of science. He did (Binford 1972, 7–8). Although Binford's published work does not contain extensive discussions of philosophical issues, the effect upon him was profound, and he profoundly influenced other archaeologists. The most explicit expression of the effect of philosophy of science upon archaeology was given in Watson, LeBlanc, and Redman's widely used text, *Explanation in Archaeology* (1971, 1984). This book relies heavily on Hempel's work on explanation, especially his *Philosophy of Natural Science* (1966) as well as "Aspects of Scientific Explanation." It is hardly an exaggeration to say that, for these authors, the hallmark of genuine science is the use of hypothetico-deductive confirmation and D-N explanation. They urge their students and colleagues to adopt this kind of methodology. The New Archaeology, which had its heyday in the late 1960s and throughout the 1970s, provides an example of an outstanding influence of philosophy of science upon an actual science. It is still a strong movement. Whether the influence was good or bad is a matter of some controversy. A detailed account of the relationships between the New Archaeology and philosophy of science is provided by Merrilee H. Salmon in *Philosophy and Archaeology* (1982).

1.3 Teleology and Functional Explanation

As we have seen, logical empiricism arose in a philosophical context in which teleology, final causes, purposes, and ends played vital roles. Those scientific philosophers who—like Braithwaite, Hempel, and Nagel—had not abandoned the notion that science can provide legitimate explanations of various kinds of phenomena were deeply concerned to provide accounts that would require admissible scientific explanations to have empirical content. They strove to prohibit pseudo-explanations that appeal to entelechies or final causes.

Nevertheless, there appeared to be cases in the empirical sciences and in common sense of bona fide explanations that are framed in terms of ends or goals. One often explains one's own behavior in this way. Why did I go to the Groceria Italiana? To get fresh pasta. In cases of this sort, however, it is not the future state of procuring pasta that explains my act; it is my preceding desire for the pasta along with the concurrent belief that that particular grocery store was a good place to get it. The explanation would be correct even if I failed to get the pasta because they were all sold out.

Midway through the first decade, Braithwaite addressed the problem of teleological explanation (1953, chap. X).[24] He saw no problem in teleological explanations of actions that are the result of conscious intention, but he also pointed out that there are cases of goal-directed behavior in which *conscious* intent seems to be absent. Among the examples Braithwaite mentions are the behavior of rats in mazes and the operation of homing torpedoes. The basic philosophical problem in these cases, given the absence of conscious intent (or any kind of intent on the part of the torpedo), involves the question of whether a future state can legitimately be said to explain a present fact.

Braithwaite notices two striking features of goal-directed behavior, *plasticity* and *variancy*. Plasticity refers to the fact that, in many cases, the agent has more than one way to achieve the goal; if one means is frustrated another will be tried. An animal trying to arrive at a place where food is located will try different routes if the usual one is blocked. Variancy refers to the fact that the goal can successfully be reached under a wide variety of initial and background conditions. A homing torpedo can—within limits—find its target regardless of the evasive action taken and—again, within limits—regardless of weather or conditions of the sea.

It has been pointed out that discussions of teleological or functional explanation often employ either or both of two models: (1) *goal-directed* behavior, as illustrated by the foregoing examples of the homing torpedo and the rat, and (2)
self-regulating behavior, as illustrated by a thermostatically controlled heat-pump
or maintenance of body temperature by a human or other animal (Canfield 1966,
Scheffler 1958). A thermostatically controlled heat-pump, for example, will keep
the temperature of a house within a certain temperature range whether the initial
temperature is below or above that range by a little or a lot. It will do so under
a wide range of outdoor temperatures, wind velocities, conditions of precipitation, etc. The human body also is capable of maintaining an internal body temperature within a narrow normal range under a wide variety of external circumstances.

In some cases, such as the homing torpedo and the heat-pump, we can give
a full causal account of the apparently purposeful behavior. In an influential article, Rosenblueth, Wiener, and Bigelow (1943) gave a cybernetic explanation of
the behavior of the homing torpedo in terms of negative feedback, and they suggested that the apparently goal-directed behavior of animals can be understood
in similar terms. These are Braithwaite's kind of example as well. Ernest Nagel,
who dealt mainly with teleological or functional explanation in physiology, focused much more intensively on the *homeostatic* or self-regulating type of example (1956; 1961, chap. 12). Just as we can provide a complete causal account of
the way in which the heat-pump controls the house temperature, starting with any
set of background conditions within its range of operation, so also, he maintained,
could we give a complete causal account of the way in which the human body
regulates its internal temperature.

According to Braithwaite, if we give a teleological explanation of the operation of a homing torpedo in terms of the goal of reaching its target, or the heat-
pump/thermostat system in terms of the goal of maintaining a given house temperature, that is merely a shorthand for a much more complicated causal explanation, and philosophically it is not very interesting. In other cases, such as the rat
in the maze, we may not be able to give a complete causal explanation in physico-
chemical terms, without reference to goal-seeking. In that case a teleological explanation is appropriate, legitimate, and interesting. But we have no reason to believe that the causal explanation in purely physico-chemical terms is impossible
in principle. Indeed, Nagel argued the stronger point that, in many physiological
cases, such causal explanations are already in hand—and that, in such cases, any
teleological explanation can, in principle, be replaced by an equivalent causal explanation. Thus, we have no basis for supposing that in any instance of goal-
directed behavior is a present fact ultimately explainable only in terms of a future
result.

If we stretch the temporal bound just a little beyond the end of the first decade,
we can include two other significant contributions to the discussion of teleological

or functional explanation – namely, the paper by Israel Scheffler (1958)[25] and a characteristically clear treatment by Hempel (1959). Hempel's discussion brings the D-N model to bear directly on the problem of functional explanation.

Hempel's philosophical concern is rather different from Braithwaite's; he does not focus particularly on the issue of final causation – on the problem of explaining present facts in terms of future goals. Given his later suggestion that subsequent events might sometimes explain earlier events (Hempel 1965, 353–54), that issue might prove embarrassing. He has, instead, a logical point in mind. The kinds of cases in which he is interested are often referred to as *functional explanations*; let us consider some simple examples. In physiology, for instance, the presence of a particular component in a given species of organisms is explained in terms of its ability to perform a function that is indispensable to the continued life or health of organisms of that sort. In humans, the blood contains hemoglobin to convey oxygen from the lungs to the other parts of the body. Without this transport of oxygen a human being could not survive.

In evolutionary biology, for another example, a feature of a species of animal is explained in terms of its enhancement of the chances of survival and reproduction. The large ears of the jackrabbit, which inhabits very hot regions, enable the animal to control its body temperature. When its body temperature rises too high, the animal seeks shade and dilates the many blood vessels in the ears. Blood coming from other parts of the body brings heat which is radiated into the environment, thereby cooling the animal. I have heard that the ears of elephants function in much the same way. In biological evolution adaptation to the environment plays a crucial role, and this appears to involve the attainment of goals.

Anthropology and sociology provide many additional cases of functional explanation. In the study of primitive societies, for instance, such prominent anthropologists as A. R. Radcliffe-Brown (1952) and B. Malinowski (1954) maintain that many – if not all – institutions, customs, and rituals are to be explained on the basis of the social functions they fulfill. The performance of a rain dance, for example, may provide an occasion for social interaction that contributes to the cohesiveness of the society. The influential sociologist R. K. Merton (1950, 1957) also advocates functional analysis in the study of human institutions.[26] He distinguishes carefully between *latent function* and *manifest function*. The rain dance has the manifest function of bringing rainfall, and it may be altogether unsuccessful in fulfilling that function; it has the latent function, however, of promoting social cohesiveness in times of distress, and it may fulfill that function quite effectively. In such cases, the latent function *explains* the survival of a practice that fails miserably to fulfill its manifest function.

Freudian psychoanalysis (which may or may not deserve to be classified as a science) is another rich source of functional explanations. For example, Freud claims that dreams serve as wish-fulfillments to prevent interruption of sleep, which is essential to the health of any human being. He also offers functional explanations of slips of the tongue or pen, and for the presence of neurotic symptoms.

The statements involved in the foregoing accounts — with the possible exception of psychoanalysis — are empirically testable and scientifically legitimate. Even if one denies that psychoanalysis is a genuine science, the basis for that judgment should not be the presence of functional explanations; the crucial issue lies in the empirical testability of its claims. It appears, then, that we have a number of different sciences that make legitimate use of functional explanations. The problem, as far as Hempel is concerned, is that functional explanations do not fit any of the logical patterns for scientific explanation, including those for inductive or statistical explanation, recognized by the received view.[27] It is *very* significant that Hempel entitled his article "The Logic of Functional Analysis" rather than "The Logic of Functional Explanation."

The fundamental problem Hempel faces can easily be seen by schematizing a simple case. He chooses the example

(3.1) The heartbeat in vertebrates has the function of circulating blood through the organism. (1959, 305)

In attempting to understand the meaning of such a statement we might be tempted simply to substitute the word "effect" for the word "function," since the circulation of blood is an effect of the beating of the heart. But the beating of the heart also has other effects that we would be unwilling to consider functions — for example, it has the effect of producing heart sounds, but that is not one of its functions. The general idea is that a function must be important to the health or survival of the organism. To do more justice to the significance of (3.1) he reformulates it as

(3.3) The heartbeat has the effect of circulating the blood, and this ensures the satisfaction of certain conditions (supply of nutriment and removal of waste) which are necessary for the proper working of the organism. (1959, 305)

To formulate the situation more abstractly, Hempel offers the following schema:

(3.4) *Basic pattern of a functional analysis*: The object of the analysis is some "item" i, which is a relatively persistent trait or disposition (e.g., the beating of the heart) occurring in a system s, (e.g., the body of a living vertebrate); and the analysis aims to show that s, is in a state, or internal condition, c_i and in an environment representing certain external conditions c_e such that under conditions c_i and c_e (jointly referred to as c) the trait i has effects which satisfy some "need" or "functional requirement" of s, i.e., a condition n which is necessary for the system's remaining in adequate, or effective, or proper, working order. (1959, 306)

It is clear from this characterization that Hempel, like Nagel, is thinking primarily in terms of the furnace analogy rather than that of the homing torpedo.

Given the fact that Hempel regards explanations as arguments, he seeks an argument that, under the conditions outlined in (3.4), would constitute an explanation of item i (the heartbeat). The crux of the problem that arises is this. We can assert (1) that if i is present and the system is in normal conditions c then n (it will be operating normally). We see (2) that n obtains (it is operating normally in these conditions). If, however, we attempt to deduce the presence of i from premises (1) and (2), we will be guilty of committing the fallacy of affirming the consequent. Given i in the circumstances we can deduce the normal operation n, but given the normal operation n we cannot validly deduce the presence of i. Moreover, Hempel recognizes, there is no inductive argument with (1) and (2) as premises which establishes the presence of i with high inductive probability; thus it cannot qualify as an acceptable I-S explanation either.

The problem to which Hempel points arises quite generally with functional explanations; it is the problem of *functional equivalents*. When we identify some item as fulfilling a function, we recognize that it is sufficient to produce some result in a certain situation. But usually we cannot claim that it is the only possible device that would fulfill that function. It is not necessary for the realization of the goal. As an obvious example, consider the large ears of the jackrabbit. This is one effective mechanism for cooling the animal, but other animals use other mechanisms. Humans perspire and dogs pant. These alternative devices are sufficient to bring about the result in question. Given only that the jackrabbit is an animal that inhabits hot regions, and thus must have some mechanism for reducing body heat, it does not follow deductively, or with high inductive probability, that the jackrabbit has extra-large ears.

Similar remarks apply to the other examples we have considered. Given that a particular ceremony enhances social cohesiveness in a given primitive society, it does not follow that no other ceremony could achieve the same end. Given that a particular dream fulfills a particular unconscious wish, it does not follow that no other dream could have done the job. Given that hemoglobin transports oxygen from one part of the body to another, it does not follow that no other chemical substance could fulfill the same function. Given that the heartbeat causes the blood to circulate, it does not follow that there is no other way for that result to come about. The problem about functional explanation can be stated simply in the following terms. In a correct D-N explanation the explanans is logically sufficient for the explanandum. In the typical functional explanation the explanandum is, given the conditions, sufficient for the explanans. From Hempel's standpoint that is just the wrong way around.

In his discussion of physiological examples, Nagel attempts to avoid the problem of functional equivalents by arguing that, if we specify with sufficient precision the kind of organism we are dealing with, only one kind of mechanism will do the trick. To cite Hempel's example, perhaps there are, in principle, mechanisms other than a beating heart that could circulate blood. Nevertheless, given the stage of evolution of homo sapiens, that is the only mechanism available. Hence if we have a healthy human—not someone undergoing surgery with a heart-lung machine attached—we know that blood is circulating, and if the blood is circulating, we know that the heart is beating. Nagel's prime example is the presence of chlorophyll in green plants (1961, 403–6). To take my example of the jackrabbit's ears, it could be claimed that in the circumstances in which that trait evolved, perspiring and panting were not biologically available. In a hot *dry* climate—the habitat of the jackrabbit— the conservation of water is absolutely critical; both perspiration and panting deplete the animal's supply of water. Whether other heat regulatory mechanisms might be available I do not know.

In view of this treatment of the problem of functional equivalents, Nagel has offered a solution to Hempel's problem about the failure of functional explanations to fit the received models. He can say quite straightforwardly that they do fit the deductive model (Nagel 1961, 405). In 1956 Nagel seemed to claim that Merton's brand of functional analysis would fit the same pattern as functional explanation in physiology, but in 1961 he had doubts that the problem of functional equivalents could be that easily dismissed in the realm of social sciences (Nagel 1961, 533–35). It seems to me, therefore, that Nagel has not succeeded in eliminating the problem of functional equivalents for teleological or functional explanation in general.

Hempel's analysis of functions is, I think, logically impeccable. If an admissible explanation of any fact must be an argument to the effect that the fact-to-be-explained *was to be expected* by virtue of the explanatory facts, then functional 'explanations' are not admissible explanations. But I have often noticed that, in philosophy as well as other human endeavors, one person's counterexample is another's modus ponens. Hempel concludes from his discussion that functional analysis cannot qualify as an admissible type of explanation; at best, it has heuristic value. Others, myself included, would take the moral to be that, since functional explanations play a legitimate scientific role, explanations cannot always be arguments of the sorts endorsed by the received view.[28] I considered it a virtue of the statistical-relevance model that it did not encounter similar problems in connection with functional explanation (W. Salmon 1982).

It turns out that there is a deep connection between the problem of functional or teleological explanation and the problem of explanation in history. Explanations in human history, as well as in the (other?) sciences of human behavior, make frequent appeals to conscious purposes and goals. As Braithwaite pointed out, such cases pose no serious problem for the philosophical theory of scientific explanation.[29] Explanations in these disciplines may, in other cases, make use of unconscious purposes, as in the example of the rain dance. In evolutionary biology functional considerations play a crucial role, and—since the time of Darwin

—it has been appropriate to deny that such appeals to functions involve the conscious purposes of a creator, or any other sort of final causation. The basic idea is that we can understand the evolutionary process in terms of efficient causes in conjunction, perhaps, with chance occurrences. Evolutionary biology thus requires a causal explication of function.

Furthermore, many authors who deal with explanation in human history insist that there is a *narrative* or *genetic* type of explanation that consists in telling the story leading up to the event to be explained.[30] Since the mere recital of just any set of preceding occurrences may have no explanatory value whatever, the narrative must involve events that are causally relevant to the explanandum if it is to serve as an explanation. This, again, demands some form of causal explanation. In human history, as in evolutionary biology, we need an account of what has come to be known as *etiological explanation* that will encompass both genetic and functional explanation.[31] In my opinion, the real breakthrough on functional explanation was provided by Larry Wright (1976) when he advanced an explicitly causal account of functional or teleological explanation.[32] Nagel took the subject up again in 1977. We shall return to this topic in §3.8.

The Second Decade (1958–67)
Manifest Destiny—Expansion and Conflict

As we have already remarked, the first decade after the publication of the Hempel-Oppenheim paper saw little published criticism—or acknowledgment—of it. Quite possibly this portion of the received view—the box in the upper left corner of Table 1—was accepted with considerable satisfaction for the most part by the philosophy of science community. The situation changed rather dramatically around 1958. This was the year in which the second volume of *Minnesota Studies in the Philosophy of Science* (Feigl et al. 1958) was published, containing Scriven's first article attacking the D-N model. Hanson's *Patterns of Discovery* (1958) appeared during the same year.

In the next few years following 1958 a great many papers on scientific explanation appeared, devoted mainly to issues we have already mentioned. These included, for example, debates on the covering law conception and on the explanation/prediction symmetry thesis. The critiques of the Hempel-Oppenheim account fall into three main categories. First, as we have seen, the Eberle-Kaplan-Montague (1961) critique and replies to it come under the heading of sympathetic efforts to find and eliminate any technical flaws in that explication. Unlike the following two types, these were critiques of the formal explication given in Part III of the Hempel-Oppenheim paper, rather than objections to the preliminary conditions of adequacy advanced in Part I. Second, as we shall see in §2.1, the attacks by Hanson, Scriven, and others were motivated by deep philosophical disagreements with anything resembling the logical empiricist point of view. Third, there were constructive efforts by philosophers such as Bromberger (1962, 1963, 1966) and Scheffler (1957, 1963) who generally accepted something like the received view, and sought ways to improve and perfect it. As we shall see in §2.3, many of the problems raised under the latter two headings were formulated with the aid of putative counterexamples that have since become standard in the literature.

During this time there were also attempts to further elaborate or defend the received view. One major effort in that direction can be found in May Brodbeck's (1962) defense of deductive explanation in her contribution to the third volume in the *Minnesota Studies* series. More significant still is Ernest Nagel's magnum opus, *The Structure of Science* (1961) which contains a great wealth of factual material on explanation in many different branches of science—physical, biological, and social—as well as in history. Nagel begins with a series of examples that include explanations in mathematics, explanations of laws (universal and statistical), deductive explanations of particular facts, probabilistic explanations, functional explanations, and genetic explanations. He also provides searching discussions of such issues as the nature of laws and the status of scientific theories. Although he does call attention to the importance of probabilistic explanation (1961, 22–23), and does discuss examples of probabilistic explanation in history (1961, 550–63), he does not provide an analysis of it. That task was not undertaken in a serious way until Hempel did so in 1962.

This period was a time of intense activity in philosophy of science in America. The Minnesota Center for the Philosophy of Science, under the direction of Herbert Feigl, had been founded in 1953, but had devoted most of its attention during the first few years to problems in the foundations of psychology (see Feigl and Scriven 1956). In the years just before 1958 the topics of discussion became more general (see Feigl et al. 1958). In 1959 Feigl and Grover Maxwell organized an ambitious program for Section L (History and Philosophy of Science) of the American Association for the Advancement of Science, the proceedings of which were published in (Feigl and Maxwell 1961). The Center for the Philosophy of Science at the University of Pittsburgh, was founded in 1960, under the direction of Adolf Grünbaum, and in the same year the Department of History & Philosophy of Science at Indiana University was created with Hanson as Chair. In addition, Robert S. Cohen and Marx Wartofsky of Boston University initiated the Boston Colloquium for the Philosophy of Science in 1960. These entities are still functioning actively. During the two academic years 1961–63, the University of Delaware conducted its Philosophy of Science Seminar (see Baumrin 1963). A great deal of important work in philosophy of science around this time was sponsored by these institutions and was published in preceedings of one kind or another. Scientific explanation was by no means the only topic—nor even the main topic, for the most part—but it was the focus of considerable attention.

Perhaps the most momentous development of the second decade was the clear articulation of theories of statistical explanation. Although a number of authors had already called attention to the need for a model of explanation in which the explanandum followed from the explanans with something less than complete deductive certainty, no real theory of that sort of explanation existed before 1962.[1] Hempel made the first attempt in that year (Hempel 1962), but followed it up with a greatly improved account in (Hempel 1965a, §3, 376–411). Much of the work of the following two decades leans heavily upon this achievement.

2.1 A Major Source of Conflict

Although Hanson was an American by birth and Scriven an Australian, both received their advanced training in England. Ludwig Wittgenstein's *Philosophical Investigations* had been published in 1953, and his influence at Cambridge and Oxford was formidable. During that era a strong opposition developed between the Wittgensteinians, who practiced ordinary language analysis, and the logical empiricists, who might be characterized as artificial language analysts. The Hempel-Oppenheim article is an outstanding example of the use of an artificial formal language for purposes of explicating a fundamental scientific concept. After Hanson and Scriven moved to America, Stephen Toulmin, who was English by birth, was the best known philosopher of science of the ordinary language school remaining in England.[2] Hanson, Scriven, and Toulmin were the most conspicuous Wittgensteinian opponents of the received view of scientific explanation during the second decade.

The basic reason for the opposition between these two schools can be seen quite readily. There is a widely accepted tripartite classification of domains in the study of languages, natural or artificial. *Syntactics* is the area in which we study merely the relationships among symbols, without regard for their relationships either to the users of the symbols or to the objects to which the symbols refer. In ordinary English it is a syntactic rule, I believe, that every sentence must begin with a capital letter and have a dot at the end (recalling that both the exclamation mark and the question mark contain dots). In a formal language, the rules for forming well-formed-formulas (wffs) are part of syntactics. *Semantics* is concerned with the relationships between symbols and the objects to which they refer, but without taking language users into account. Among its important concepts are designation, meaning, and truth. Deductive validity can be construed either syntactically or semantically; for purposes of our discussion it does not matter much which interpretation is chosen. *Pragmatics* takes account of the users of symbols, as well as the interrelations among the symbols and the relationships between the symbols and the entities they stand for. It emphasizes, among many other things, the context in which a statement is made, the purpose of the person who makes it, and the presuppositions that are shared in that context.

The explications offered by the logical empiricists were usually constructed entirely in syntactical and/or semantical terms. The point is well illustrated by the Hempel-Oppenheim treatment of scientific explanation. They start with a standard first order logic, which can be characterized in purely syntactic terms, and proceed to offer semantic rules for its interpretation. All of the key notions—e.g., lawlike sentence, law, theory, explanans—are defined semantically. Pragmatics plays hardly any role at all. Writing in the third decade, Nicholas Rescher (1970, 6–8) points to the many pragmatic features of scientific explanation, but maintains that, for purposes of logical analysis, it is best to abstract from them.

For anyone who focuses primarily on the ordinary uses of language, the pragmatic aspects will be most conspicuous. Some explanations, for example, may be requested by formulating why-questions. The answers are to be judged in terms of the interests and background knowledge of the questioner. An explanation of a particular phenomenon that is entirely satisfactory for one person may be totally inappropriate for another.

It is easy to see why deep philosophical conflict arose in that situation. The ordinary language philosopher finds the logical empiricist insensitive to human needs and interests. The logical empiricist finds the ordinary language philosopher unappreciative of the objective features that determine whether a proffered explanation is a bona fide scientific explanation. The logical empiricists employed formal techniques; the ordinary language philosophers tended to deprecate and avoid them. In those days formal pragmatics did not exist in any highly developed form. As we shall see in the fourth decade, formal pragmatics now plays a crucial role in various approaches to explanation, especially that of Bas van Fraassen.

Hempel was aware from the beginning that when scientists and others offer explanations they often omit parts that are obvious. Often the omitted part is the law. Such explanations are unobjectionable to Hempel if the law is obvious to both the questioner and the respondent; it would be needlessly pedantic to insist on mention of the obvious. Partial explanations and explanation sketches are frequently acceptable. To this degree, at least, Hempel acknowledges the pragmatic aspects of explanation (see 1965a, §4–5, regarding these and other pragmatic considerations). Nevertheless, Hempel insists, when you spell out the correct explanation in complete detail, it will always contain a law.

Scriven takes an opposite attitude. He notices that in many—if not most—cases, an explanation of one particular fact consists in citing another particular fact. Why did the automobile radiator rupture? Because the car was left outside overnight without antifreeze, and the temperature fell to 10° F during the night. If this explanation is questioned, the law concerning the expansion of water upon freezing may be cited, but, according to Scriven, not as a suppressed part of an incomplete explanation. The explanation, as given, was satisfactory. If the law is invoked it serves to justify the explanation, not to complete it.

This view is in strict accord with a thesis about laws that had been shared by some early logical positivists and the ordinary language philosophers, namely, that laws are not statements of fact, but rather, rules of inference. On that view, the sentence "Silver is an electrical conductor" is not a true generalization; it is an 'inference ticket' that entitles one to conclude from "This object is made of silver" that "This object is capable of conducting electricity." Hempel and Oppenheim insisted that a correct D-N explanation be a valid deductive argument, but they did not suggest that the argument include a statement of the rules of deduction to which it conforms. If anyone questions the validity of a given explanation, we can trot out the rules to demonstrate its validity. Philosophers who, like Toulmin, regarded laws as 'inference tickets,' likewise objected to the idea that these rules of inference be included as parts of the explanation itself. They stand on the side, so to speak, to be called up if the correctness of a given explanation is challenged. In making this point about explanation, Scriven refers to the *role-justifying function* of laws (1962, 200, 207, 211).

2.2 Deeper Linguistic Challenges

Not long after Hanson and Scriven had fired their opening salvos against the Hempel-Oppenheim approach, a rather different sort of linguistic approach was initiated by Sylvain Bromberger. Informed by empirical linguistics, which at the time had had virtually no impact on ordinary language philosophers, Bromberger advanced a much more detailed and precise account of explanation than was offered by any of them.

At the beginning of his first paper, "An Approach to Explanation," Bromberger (1962, 72–73) invites consideration of three statements that involve explanation:

(1) The kinetic theory of gases explains why the state equation of vapours near condensation differs markedly from that of an ideal gas.

(2) Someone explained to somebody how World War II might have been avoided.

(3) Newton explained a long time ago the variations of the tides.

He selects statements like (2) for primary attention as a matter of convenience, he says, and he readily extends his considerations to statements like (3). It should be noted, however, that when he gets to the end of this essay he has little to say about statements like (1), and he admits to having no theory that adequately handles them. At the close of the essay he remarks, "The account of the nature of explanation just given falls short of what is eventually wanted: it fails to provide the sort of insight that can be translated into explicit standards and into a patttern of analysis applicable to all explanations and capable of deciding their correctness; it fails to make explicit the criteria that make correct explanations *correct* explanations" (1962, 104–5). So I think the choice of statements like (2)—at least as opposed to those like (1)—is based on more than convenience. What Bromberger admittedly failed to achieve is just the sort of thing Hempel and Oppenheim were trying to accomplish.

From a linguistic standpoint, a statement like (2) is especially tractable because it implies the occurrence of a particular sort of linguistic performance involving two particular individuals or groups of people at some particular time. In addition to the people and the time, such statements incorporate some form of the verb "to explain," and 'something' that is explained, "where the 'something' can be specified by means of an indirect question" (1962, 73). It is close to what Bas van

Fraassen, in the fourth decade, refers to as *the topic* of a question. More precisely, Bromberger is concerned with statements of the form "A E to B W," where "A" and "B" indicate places that can be occupied by terms designating individuals or groups of people, "E" indicates a place where a tensed form of "to explain" occurs, and "W" indicates a place where an indirect question occurs. The heavily linguistic character of this approach is shown by the fact that this characterization could not even purport to apply to explanations in any language other than English. The Hempel-Oppenheim model, by contrast, could apply to explanations offered in myriad different languages.

Bromberger's linguistic analysis proceeds by classifying the verb "to explain," according to a taxonomy provided by Zeno Vendler, as an *accomplishment* term. Other verbs fall into the categories of *activity* terms, *state* terms, or *achievement* terms. Activity terms and accomplishment terms apply to doing something that occupies a span of time, but an accomplishment term refers to an activity that can result in a completion. Explaining and reading are activities that take place in that way, but when an activity term is used in the simple past, it does not imply any completion. If I say I read last night, that does not mean I finished a story, an article, or a book. It is an activity in which I could be engaged and which I could have stopped doing without finishing anything. It means merely that I spent some time engaged in that activity. If I say that I explained in class yesterday, that claim would not be correct unless I had completed an explanation. If no explanation was completed, I could correctly say that I was trying to explain, or that I got part way through an explanation, but not that I explained anything.

The difference between activity terms and accomplishment terms is readily seen when we compare their simple past tense. Both types have a simple past tense which implies that the continuous present was applicable at some moments in the past. The simple past tense of an activity term is applicable as soon as such moments have passed, and implies only the existence of such moments in the past. Aristotle walked. This implies that during some moments in the past Aristotle was walking. It does not tell whether or not Aristotle is through walking. The simple past tense of accomplishment terms implies more. It implies that relevant activities took place in the past, but furthermore that they have come to an end. And not to a mere stop, but to a conclusion. In other words, the simple past tense of accomplishment verbs entails that something has been finished, completed, that might, in principle, have been left unfinished, incomplete, that might have stopped before coming to its logical end. (1962, 75)

State terms, such as "to know" and "to love" (in their non-episodic senses), truly apply in the simple present tense to individuals at different times even though the individual is in that state just once—e.g., he loves her, said at any time throughout their 35 years of marriage. Achievement terms can be truly applied in the simple present at more than one time only if the individual achieves more than once. She wins at tennis at time t_1 and she wins at tennis at t_2 (where t_1 is different from t_2) only if she wins two different games (or sets, or matches).

Since "to explain" is an accomplishment term, "A good analysis should therefore make explicit the nature of the completion implied by statements in 'A E to B W' form in which 'to explain' occurs in the *simple past tense*; it should bring out what must be the case for a statement in 'A explain*ed* to B W' form to be true' " (1962, 76).

Another item of important concern about statements of the form 'A E to B W' is the nature of the indirect question at the place indicated by "W." Clearly, Bromberger emphasizes, certain sorts of questions are appropriate, while other kinds are inappropriate.

Many kinds of indirect question can occupy the position indicated by 'W', and they may open on a variety of interrogatives[3] – 'Why', 'How', 'Whence', 'Whither', 'What' – but not *every* indirect question is at home there; some would be out of place, awkward, reminiscent of Eisenhower prose, *e.g.* 'what the distance in miles between London and Paris is' or 'whether it will rain tomorrow' or 'what time it is' or 'which sister Sam married'. A good analysis should show why some indirect questions do not sit well in these contexts. (1962, 74)

Bromberger invites us to consider two questions:

(A) What is the height of Mt. Kilimanjaro?
(B) Why do tea kettles emit a humming noise just before the water begins to boil? (1962, 80)

Each of these is a sound question – each has a correct answer and neither has any false presuppositions.

Although he does not know a correct answer for either, he knows quite a bit about what would constitute a correct answer. A correct answer to (A) might be a positive integer followed by an expression, such as "feet," designating a unit of length. If the answer is given in feet, he knows that the number will be greater than 100 and less than 30,000; if some other unit is used, it will be possible to convert the answer into feet, and the same numerical limits will apply. He knows enough about the answer to be able to exclude many expressions – e.g., "12 feet," "Morton White," "19,321 pounds." Although he does not know a correct answer, there is a straightforward sense in which he has thought of a correct answer, for he could write down and understand any integral numeral between "100" and "30,000."

A correct answer for (B) would be statable as a sentence (which may be a conjunction of sentences) following the word "because." This sentence would include mention of something that happens whenever water is just about to boil, and it would include mention of something that creates vibrations of air of suitable frequency and amplitude. Other requirements on an answer might be offered, and they would serve to exclude many expressions as possible answers, but altogether they do not add up to a correct answer. In the case of (B), however, unlike (A), he has not thought of any answer that could possibly be a correct answer.

This is a key feature of explanation. Someone could tell Bromberger the height of Kilimanjaro, but it would be incorrect usage to say anyone had explained it to him. Perhaps there is a geophysical explanation of the height of Kilimanjaro, but question (A) was not a request for it. Question (B) was a request for an explanation. He recapitulates:

(i)   I take both (A) and (B) to be sound questions, to admit of a right answer.
(ii) I know, or believe I know, enough about each answer to be able to eliminate a number of possible utterances, *i.e.* of expressions, as not being formulations of it.
(iii) In the case of (A) I can think of some possible utterances that I cannot eliminate in this way.
(iv) In the case of (B) I can think[4] of no expression that I cannot eliminate in this way. (1962, 82)

On the basis of these considerations, Bromberger defines the notion of a *p-predicament* as follows:

S is in a p-predicament with regard to question Q if and only if, on S's views, Q admits of a right answer, but S can think of no answer to which, on S's views, there are no decisive objections. (1962, 82)

He points out that one could not be in a p-predicament with regard to an indirect question beginning with "whether" that requires a "yes" or "no" answer — e.g., whether it is raining here just now. Even if one does not know the correct answer, it surely has been thought of. Similarly, for reasons that have already been mentioned, one would not be in a p-predicament regarding indirect questions like (A), beginning with "what is," that call for quantitative answers. Likewise, one cannot be in a p-predicament with respect to indirect questions, beginning with "which," that call for a selection from a well-defined set of alternatives. In my present state of knowledge I may not know which planet in the solar system has the highest surface temperature, but I believe I have thought of the possible candidates, and I suppose it is either Mercury or Venus.

To arrive at a characterization of explanation, Bromberger also defines the notion of a b-predicament:

S is in a b-predicament with regard to Q if and only if the question mentioned in it admits of a right answer, but that answer is beyond what the person men-

tioned can conceive, can think of, can imagine, i.e. is something that that per-
son cannot remember, cannot excogitate, cannot compose. (1962, 90, italics
not in original)

The main difference between a p-predicament and a b-predicament is that the latter refers to a right answer, whereas the former refers to possible answers that
cannot be eliminated.

Bromberger offers a series of four hypotheses concerning the nature of explanations, and he rejects the first three. In stating these hypotheses he refers to the
person or persons doing the explaining as the *tutor* and to the recipient as the *tutee*. We shall look only at the fourth hypothesis:

The essential characteristics of explaining episodes are the following:

(a) the question is sound, i.e. admits of a right answer;

(b) the tutor is rational and knows the right answer to the question at the
time of the episode;

(c) during the episode the tutor knows, or believes, or at least assumes that
at the beginning of the episode the tutee was in a p-predicament with regard
to the question,

or that, at the beginning of the episode the tutee was in a b-predicament with
regard to the question,

or that, at the beginning of the episode the tutee was in either a p-
predicament or a b-predicament with regard to the question;

(d) in the course of the episode the tutor presents the facts that, in his opinion, the tutee must learn to know the right answer to the question;

(e) in the course of the episode the tutor also provides the tutee with such
instruction as he (the tutor) thinks necessary to remove the basis of whichever
of the states mentioned in (c) he deems the tutee to be in;

(f) at the end of the episode all the facts mentioned in (d) and (e) have been
presented to the tutee by the tutor. (1962, 94–95)

Bromberger acknowledges that this lengthy characterization contains redundancies: (a) is entailed by (b); (c) and (f) are entailed by (d) and (e). Removing the
redundancies, we have

The essential characteristics of explaining episodes are the following:

(b) the tutor is rational and knows the right answer to the question at the
time of the episode;

(d) in the course of the episode the tutor presents the facts that, in his opinion, the tutee must learn to know the right answer to the question;

(e) in the course of the episode the tutor also provides the tutee with such
instruction as he (the tutor) thinks necessary to remove the basis of whichever
of the predicaments he deems the tutee to be in.

All the kinds of explanation we have discussed so far fit the schema "A E to B W," where "W" indicates a place where an indirect question occurs. After presenting this account, Bromberger goes on to discuss explanations that fit the schema "A E to B X," where "A," "B," and "E" have the same significance, but "X" is the location of something other than a question, for example, a noun phrase. He rather easily shows how these can be handled in terms of the preceding type, and remarks on the desirability of keeping clear on "the essential connection that links explaining to questions." (1962, 100)

Statement (3) of Bromberger's initial set, it will be recalled, is "Newton explained a long time ago the variations of the tides." This statement does not imply an explaining episode of the sort we have been discussing. Here "explained" is not an accomplishment term; it is, rather, an *achievement* term. The statement means that Newton solved the problem. His solution of the problem is such that, for anyone who understands it, that person can provide an explanation of the sort we previously discussed. "The connection, then, between the truth-conditions of 'A E W' and 'A E to B W' is that to have explained something in the sense now under consideration is to have become able to explain something in the [previous] sense . . . as a result of one's own endeavours and ingenuity." (1962, 101)[5]

Bromberger's first example of a statement involving "to explain" — the one about the ability of the kinetic theory to explain why certain vapors do not behave as ideal gases — has the form "T explains W," where "T" does not indicate a place for a term referring to a person or group of people. In this sentence, "to explain" functions as a state term; it refers to the continuing ability of a theory to function in explanations of the sort we have already discussed. As I remarked at the outset, Bromberger does not claim to have an adequate treatment of this kind of explanation.

I have discussed Bromberger's approach in considerable detail and I have presented lengthy direct quotations to give a clear flavor of what he is attempting, and to contrast it with the sort of thing Hempel and Oppenheim were trying to do. Clearly, Bromberger has great sensitivity to the nuances of language. He goes to considerable length to understand the usages of "to explain" — for instance, to discover the differences between explanation-seeking questions and other kinds of questions. In his principal examples the explanation-seeking question was a why-question, but he denies that all and only why-questions are explanation-seeking. His identification of p-predicaments and b-predicaments involves deep insight into the nature of questions. It also exhibits clear appreciation of many of the pragmatic aspects of explanation.

In the end, however, as Bromberger is clearly aware, his characterization of explanation employs the unanalyzed notion of a right answer to a why-question or any other kind of explanation-seeking question. As I said above, what Hempel and Oppenheim were trying to do is to analyze the notion of a correct answer to an explanation-seeking question. Although they undoubtedly made significant progress on their enterprise, they offered no criteria at all to enable us to distinguish explanation-seeking from non-explanation-seeking questions. Bromberger's paper, which he aptly titled "An Approach to Explanation" (rather than, say, "A Theory of Explanation") provides a valuable prolegomenon to the Hempel-Oppenheim project, but, as he recognized, not a substitute for it.

Toward the end of the second decade, Bromberger (1966) returned to this issue and attempted to give a partial answer. Although he still insists that not every request for an explanation can be given in the form of a why-question, he does try to characterize correct answers to explanation-seeking why-questions. He offers several counterexamples – similar in import to the flagpole (CE-2, to be discussed in the next section) – to show that not every argument that satisfies the criteria for D-N explanation, as set out by Hempel and Oppenheim, qualifies as an acceptable answer to a particular why-question. However, he does not dispute the thesis that these conditions are necessary for satisfactory answers – for satisfactory *deductive* answers, we should say, for he does not deny that inductive or statistical explanations are possible. The theory of why-questions can thus be seen as a friendly amendment to the Hempel-Oppenheim account.[6] He does not, however, discuss any of the issues surrounding the theories of statistical explanation that were published by Hempel shortly before the appearance of this article by Bromberger.

As his point of departure, Bromberger rejects the covering law conception that identifies explanation, or correct answers to why-questions, with subsumption under true lawful generalizations. Consider the question, "Why does this live oak keep its leaves during the winter?" The appropriate response, he suggests, is "All live oaks do!" "Because all live oaks do" is, he claims, not correct. Since he maintains that answers to why-questions are sentences that follow "because," the correct response is not an answer, but a rejection of the question (1966, 102). The question suggests that there is something atypical about this particular tree; the response denies that supposition.

Bromberger's basic idea is that why-questions arise "when one believes that the presupposition is true, views it as a departure from a general rule, and thinks that the conditions under which departures from the general rule occur can be generalized" (1966, 100). To develop this thesis we need several definitions.

(1) A *general rule* is simply a *lawlike* generalization of the form

$$(x)[(F_1x . F_2x . . . F_jx) \supset (S_1x . S_2x . . . S_kx)] \qquad (j, k \geq 1)$$

Examples are: All gold is malleable, no rubber bands are brittle, and the velocity of an object never changes. Obviously, a general rule need not be true or even plausible (1966, 97).

(2) Bromberger introduces the concept of a *general abnormic law*, but the definition is quite complex, and we do not need all of the details for our discussion. He remarks that every general abnormic law is equivalent to a conjunction of spe-

cial abnormic laws. It should, however, be emphatically noted that general abnormic laws are true (1966, 97–99).

(3) A *special abnormic law* is a true lawlike generalization that contains an unless-clause. Examples are: The velocity of an object does not change unless the net force on it is not equal to zero; no rubber band is brittle unless it is very old or very cold. Special abnormic laws have the following form:

$$(x)\{(F_1x \cdot F_2x \ldots F_jx) \supset [-Ex \equiv (A_1x \, A_2x \ldots A_nx)]\} \quad (j, n \geq 1)$$

which can obviously be reformulated as

$$(x)\{(F_1x \cdot F_2x \ldots F_jx) \supset [Ex \equiv -(A_1x \, A_2x \ldots A_nx)]\} \quad (j, n \geq 1)$$

They must also fulfill further restrictions that need not be spelled out for our purposes.[7]

(4) Bromberger defines the *antonymic predicates of an abnormic law,* but we need only the concept of *antonymic predicates of a special abnormic law.* These are the predicates that occur in the place of "E" and its negation in the foregoing forms. For example, "brittle" and "nonbrittle" would be the antonymic predicates in "No rubber bands are brittle unless they are old or cold" (1966, 99).

(5) Finally, we need the concept of the *completion of a general rule by an abnormic law.* An abnormic law is the completion of a general rule if and only if the general rule is false and is obtainable from the abnormic law by dropping the "unless"-clause (1966, 99).

With this terminology in place we can now characterize the concept of a correct answer to a why-question:

> Here then is the relation: *b* is the *correct answer* to the question whose presupposition is *a* if and only if (1) there is an abnormic law *L* (general or special) and *a* is an instantiation of one of *L*'s antonymic predicates; (2) *b* is a member of a set of premises that together with *L* constitute a deductive nomological explanation whose conclusion is *a*; (3) the remaining premises together with the general rule completed by *L* constitute a deduction in every respect like a deductive nomological explanation except for a false lawlike premise and false conclusion, whose conclusion is a contrary [or contradictory] of *a*. (1966, 100)[8]

Consider one of Bromberger's examples. Suppose we have a simple pendulum that is oscillating with a particular period p. It consists of a bob suspended by a string of a given length. From the law of the simple pendulum, the acceleration of gravity at the earth's surface g, and the length l of the string, one can deduce the period p of our pendulum. This argument would qualify as a deductive nomological explanation according to the Hempel-Oppenheim account, and it would be regarded as a correct explanation of the period. However, from the same law, the same value for the acceleration of gravity, and the period of the pendulum we can deduce the length of the string. This would not seem to constitute a good answer to the question of why the string has its particular length l.

Consider the general rule, "No simple pendulum oscillates with period p." It is, of course, false. It can, however, be completed by the abnormic law, "No simple pendulum has the period p unless[9] it has a suspension of length l." From the abnormic law and the fact that the suspension has length l, we can deduce that it has period p. From the general rule and the fact that we have a simple pendulum we can deduce that it is not oscillating with period p. Thus, it is correct, according to Bromberger's account, to answer the why-question about the period of oscillation with facts about the length of the suspension and the local acceleration of gravity, and the law of the simple pendulum. However, to answer the question, "Why does this pendulum have a suspension of length l?" it would be manifestly incorrect to answer, "Because it is oscillating with period p." For consider the putative abnormic law, "No pendulum has a suspension of length l unless it is oscillating with period p." This is patently false, for a pendulum that is not oscillating may have just such a suspension. Bromberger thus attempts to solve a nagging problem about asymmetries of explanation.

Bromberger closes this essay with the following remark:

> It may seem odd that abnormic laws should be associated with a special interrogative. But they are, after all, the form in which many common-sense generalizations that have been qualified through the ages are put. They are also a form of law appropriate to stages of exploratory theoretical developments when general rules are tried, then amended, until finally completely replaced. We are always at such a stage. (1966, 107)

What Bromberger has noticed, I believe, is a deep fact about scientific explanations. To provide an adequate explanation of any given fact, we need to provide information that is relevant to the occurrence of that fact — information that *makes a difference* to its occurrence. It is not sufficient simply to subsume an occurrence under a general law; it is necessary to show that it has some special characteristics that account for the features we seek to explain. As we shall see, this notion of explanatory relevance plays a key role in the development of theories of statistical explanation.

As I have already remarked, and as we shall see in greater detail, the formulation of theories of inductive or statistical explanation was one of the most significant — perhaps the most significant — development in the second decade. Being confined to deductive types of explanation, Bromberger's treatment did not encompass this broader range of issues. Moreover, as an incisive critique by Paul Teller (1974) showed, it does not even achieve its aim within its intended range of application. Nevertheless, it provided a fertile basis for subsequent work on the pragmatics of explanation, especially that of Achinstein and van Fraassen during the fourth decade. In addition, it has served as a major stimulus to Philip Kitcher in developing the novel unification approach to explanation embodied in "Explanatory Unification and the Causal Structure of the World" in this volume (Kitcher & Salmon 1989) especially (Bromberger 1963).

2.3 Famous Counterexamples to the Deductive-Nomological Model

At the conclusion of §1.1 we raised a number of general issues arising out of the Hempel-Oppenheim treatment of scientific explanation of particular occurrences. During the second decade many of them were vividly posed in terms of counterexamples that have since become quite standard. The existence of such a standard set is, in itself, a tribute to the solidity of the received view. Let us take a look at some of the best known.

One rather obvious problem has to do with the temporal relations between the explanatory facts (as expressed by the singular sentences in the explanans) and the fact-to-be-explained (as expressed by the explanandum-sentence). In the schema reproduced above (p. 13; H-O 1948, 249) the Cs are labeled as "antecedent conditions," but in the formal explication no temporal constraints are given. Indeed, no such temporal constraints are mentioned even in the informal conditions of adequacy. This issue has been posed in terms of the explanation of an eclipse.

(CE-1) The eclipse. Going along with the D-N model, we might, for example, regard a total lunar eclipse as satisfactorily explained by deducing its occurrence from the relative positions of the earth, sun, and moon at some prior time in conjunction with the laws of celestial mechanics that govern their motions. It is equally possible, however, to deduce the occurrence of the eclipse from the relative positions of the earth, sun, and moon at some time after the eclipse in conjunction with the very same laws. Yet, hardly anyone would admit that the latter deduction qualifies as an explanation.[10] One might suppose that the failure to impose temporal restrictions was merely an oversight that could be corrected later, but Hempel (1965a, 353) raises this question explicitly and declines to add any temporal constraint.[11]

Another issue, closely related to the matter of temporal priority, has to do with the role of causality in scientific explanation. Our commonsense notion of explanation seems to take it for granted that to explain some particular event is to identify its cause and, possibly, point out the causal connection. Hempel and Oppenheim seem to share this intuition, for they remark, "The type of explanation which has been considered here so far is often referred to as causal explanation" (H-O 1948, 250). In "Aspects of Scientific Explanation," while admitting that some D-N explanations are causal, Hempel explicitly denies that all are (1965a, 352–54).

The problems that arise in this connection can readily be seen by considering several additional well-known examples.

(CE-2). Bromberger's flagpole example. A vertical flagpole of a certain height stands on a flat level piece ground.[12] The sun is at a certain elevation and is shining brightly. The flagpole casts a shadow of a certain length. Given the foregoing facts about the height of the flagpole and the position of the sun, along with the law of rectilinear propagation of light, we can deduce the length of the shadow. This deduction may be accepted as a legitimate D-N explanation of the length of the shadow. Similarly, given the foregoing facts about the position of the sun and the length of the shadow, we can invoke the same law to deduce the height of the flagpole. Nevertheless, few people would be willing to concede that the height of the flagpole is explained by the length of its shadow.[13] The reason for this asymmetry seems to lie in the fact that a flagpole of a certain height causes a shadow of a given length, and thereby explains the length of the shadow, whereas the shadow does not cause the flagpole, and consequently cannot explain its height.

(CE-3) The barometer. If a sharp drop in the reading on a properly functioning barometer occurs, we can infer that there will be a storm—for the sake of argument, let us assume that there is a law that whenever the barometric pressure drops sharply a storm will occur. Nevertheless, we do not want to say that the barometric reading explains the storm, since both the drop in barometric reading and the occurrence of the storm are caused by atmospheric conditions in that region. When two different occurrences are effects of a common cause, we do not allow that either one of the effects explains the other. However, the explanation of the storm on the basis of the barometric reading fits the D-N model.

(CE-4) The moon and the tides. Long before the time of Newton, mariners were fully aware of the correlation between the position and phase of the moon and the rising and falling of the tides. They had no knowledge of the causal connection between the moon and the tides, so they had no explanation for the rising and falling of the tides, and they made no claim to any scientific explanation. To whatever extent they thought they had an explanation, it was probably that God in his goodness put the moon in the sky as a sign for the benefit of mariners. Nevertheless, given the strict law correlating the position and phase of the moon with the ebb and flow of the tides,[14] it was obviously within their power to construct D-N explanations of the behavior of the tides. It was not until Newton furnished the causal connection, however, that the tides could actually be explained.

One of the most controversial theses propounded by Hempel and Oppenheim is *the symmetry of explanation and prediction*. According to this view, the very same logical schema fits scientific explanation and scientific prediction; the sole difference between them is pragmatic. If the event described by E has already occurred, we may ask why. A D-N explanation consisting of a derivation of E from laws and antecedent conditions provides a suitable response. If, however, we are in possession of the same laws and antecedent conditions before the occurrence of E, then that same argument provides a prediction of E. Any D-N explanation is an argument that, were we in possession of it early enough, would enable us to anticipate, on a sound scientific basis, the occurrence of E. Since every D-N explanation involves laws, a hallmark of explanations of this type is that they provide *nomic expectability*. [15]

In discussing the symmetry of explanation and prediction in the preceding paragraph, I was tacitly assuming that the so-called antecedent conditions in the explanans are, in fact, earlier than the explanandum event. However, in view of Hempel's rejection of any requirement of temporal priority, the symmetry thesis must be construed a bit more broadly. Suppose, for example, that the explanandum-event E occurs before the conditions C in the explanans. Then, as I construe the symmetry thesis, we would be committed to the view that the D-N explanation is an argument that could be used subsequent to the occurrence of the explanatory conditions C to retrodict E. It is quite possible, of course, that E has occurred, but that we are ignorant of that fact. With knowledge of the appropriate laws, our subsequent knowledge of conditions C would enable us to learn that E did, in fact, obtain. Parallel remarks could be made about the case in which C and E are simultaneous. Thus, in its full generality, the symmetry thesis should be interpreted in such a way that "prediction" is construed as "inference from the known to the unknown."[16]

As Hempel later pointed out in "Aspects of Scientific Explanation," the symmetry thesis can be separated into two parts: (i) Every D-N explanation is a prediction — in the sense explained in the preceding paragraph — and (ii) every (nonstatistical) scientific prediction is a D-N explanation. It is worthwhile, I think, to distinguish a *narrower symmetry thesis*, which applies only to D-N explanations of particular facts, and a *broader symmetry thesis*, which applies to both D-N and I-S explanations of particular facts. According to the narrower thesis, every *nonstatistical* prediction is a D-N explanation; according to the broader thesis, every prediction is an explanation of either the D-N or I-S variety. Given the fact that statistical explanation is not explicated in the Hempel-Oppenheim article, only the narrower symmetry thesis is asserted there. The broader thesis, as we shall see, was advocated (with certain limitations) in "Aspects."

Nevertheless, various critics of the Hempel-Oppenheim article failed to take sufficient notice of the explicit assertion that not all legitimate scientific explanations are D-N — that some are statistical. Scriven (1959) strongly attacked subthesis (i) — that all explanations could serve as predictions under suitable pragmatic circumstances — by citing evolutionary biology and asserting that it furnishes explanations (of what has evolved) but not predictions (of what will evolve). If, as I believe, evolutionary biology is a statistical theory, then Scriven's argument applies at best to the broader, not the narrower symmetry thesis. Although this argument was published in 1959,[17] it does, I believe, pose a serious problem for the theory of statistical explanation Hempel published three years later. In the same article Scriven set forth a widely cited counterexample:

(CE-5) Syphilis and paresis. Paresis is one form of tertiary syphilis, and it can occur only in individuals who go through the primary, secondary, and latent stages of the disease without treatment with penicillin. If a subject falls victim to paresis, the explanation is that it was due to latent untreated syphilis. However, only a relatively small percentage—about 25%—of victims of latent untreated syphilis develop paresis. Hence, if a person has latent untreated syphilis, the correct prediction is that he or she will not develop paresis. This counterexample, like the argument from evolutionary biology, applies only to the broader symmetry thesis.

When the narrower symmetry thesis is spelled out carefully, it seems impossible to provide a counterexample for subthesis (i)—that every explanation is a prediction (given the right pragmatic situation). That subthesis amounts only to the assertion that the conclusion of a D-N argument follows from its premises. Against subthesis (i) of the broader symmetry thesis the syphilis/paresis counterexample is, I think, quite telling.

When we turn to subthesis (ii) of the narrower symmetry thesis—i.e., that every (nonstatistical) prediction is an explanation—the situation is quite different. Here (CE-3) and (CE-4) provide important counterexamples. From the barometric reading, the storm can be predicted, but the barometric reading does not explain the storm. From the position and phase of the moon, pre-Newtonians could predict the behavior of the tides, but they had no explanation of them. Various kinds of correlations exist that provide excellent bases for prediction, but because no suitable causal relations exist (or are known), these correlations do not furnish explanations.

There is another basis for doubting that every scientific prediction can serve, in appropriate pragmatic circumstances, as an explanation. Hempel and Oppenheim insist strongly upon the covering law character of explanations. However, it seems plausible to suppose that some respectable scientific predictions can be made without benefit of laws—i.e., some predictions are inferences from particular instances to particular instances. Suppose, for instance, that I have tried quite a number of figs from a particular tree, and have found each of them tasteless. A friend picks a fig from this tree and is about to eat it. I warn the friend, "Don't eat it; it will be tasteless." This is, to be sure, low-level science, but I do not consider it an unscientific prediction. Moreover, I do not think any genuine laws are involved in the prediction. In (1965, 376) Hempel considers the acceptability of subthesis (ii) of the symmetry thesis an open question.

There is another fundamental difficulty with Hempel and Oppenheim's explication of D-N explanation; this one has to do with explanatory relevance. It can be illustrated by a few well-known counterexamples.

(CE-6) The hexed salt. A sample of table salt has been placed in water and it has dissolved. Why? Because a person wearing a funny hat mumbled some nonsense syllables and waved a wand over it—i.e., cast a dissolving spell upon it. The explanation offered for the fact that it dissolved is that it was hexed, and all hexed samples of table salt dissolve when placed in water. In this example it is *not* being supposed that any actual magic occurs. All hexed table salt is water-soluble because all table salt is water-soluble. This example fulfills the requirements for D-N explanation, but it manifestly fails to be a bona fide explanation.[18]

(CE-7) Birth-control pills. John Jones (a male) has not become pregnant during the past year because he has faithfully consumed his wife's birth-control pills, and any male who regularly takes oral contraceptives will avoid becoming pregnant. Like (CE-6), this example conforms to the requirements for D-N explanation.

The problem of relevance illustrated by (CE-6) and (CE-7) is actually more acute in the realm of statistical explanation than it is in connection with D-N explanation. Insofar as D-N explanation is concerned, it is possible to block examples of the sort just considered by any of several technical devices.[19] We will return to this issue when we discuss statistical explanation.

2.4 Statistical Explanation

In an article entitled "The Stochastic Revolution and the Nature of Scientific Explanation," Nicholas Rescher (1962) made an eloquent plea for an extension of the concept of scientific explanation beyond the limits of deductive explanation. The "stochastic revolution" yields "forcible considerations . . . that militate towards a view of explanation prepared to recognize as an 'explanation' of some fact an argument which provides a rationalization of this fact from premises which render it *not necessary but merely probable*" (ibid., 200). He adds, "To refuse to accord to such explanatory reasonings the title of 'explanation' is to set up so narrow a concept of explanation that many of the reasonings ordinarily so-called in modern scientific discussions are put outside the pale of *explanations proper* by what is in the final analysis, a fiat of definition buttressed solely by fond memories of what explanation used to be in nineteenth-century physics" (ibid. 204).

The most important development to occur in the second decade (1958–67) of our chronicle—the explicit treatment of statistical explanation—had its public inception in 1962. Although Rescher clearly recognized the inductive character of such explanations, neither he nor any of several other authors who recognized the legitimacy of statistical explanation offered an explicit model. They thought,

I suspect, that nothing was needed beyond a trivial relaxation of the requirements imposed on D-N explanations. Hempel appears to have been the first to notice the profound difficulties involved in statistical explanation, in particular, the problem of *ambiguity*, over which he labored long and hard. The first explicit model was offered in his "Deductive-Nomological vs. Statistical Explanation" (1962) in volume III of *Minnesota Studies*. It constitutes the first serious attempt (of which I am aware) by any philosopher or scientist to offer a detailed and systematic account of any pattern of statistical explanation.[20] In this paper, Hempel deals with statistical explanations of particular facts – the lower left-hand box of Table 1 – and characterizes what he later calls the *inductive-statistical (I-S)* model. In the same year (1962a) he offered a brief popular account of both deductive and inductive explanations of particular facts in history as well as in science. In section 3 of "Aspects," he offers a much improved account of the I-S model, and also introduces the *deductive-statistical (D-S)* model.

2.4.1 The Deductive-Statistical Model

In a D-S explanation, a statistical law[21] is explained by deriving it from other laws, at least one of which is statistical. There is no prohibition against the explanans containing universal laws as well. A time-honored example will illustrate how this sort of explanation goes.

In the 17th century the Chevalier de Méré posed the following question: Why is it that, when a player tosses a pair of standard dice in the standard manner 24 times, he has less than a 50–50 chance of getting double 6 at least once? The answer is as follows. On each toss of a fair die, the chance of getting 6 is 1/6. This probability is independent of the outcome of the toss of the other die. Consequently, the probability of double 6 on any throw is 1/36, and the probability of not getting double 6 is 35/36. The outcome of any toss of the pair of dice is independent of the outcomes of previous tosses; hence, the probability of not getting double 6 on n throws is $(35/36)^n$. When we calculate the value of $(35/36)^{24}$ we find it is greater than 1/2; therefore, the probability of getting double 6 in 24 tosses is less than 1/2. Assuming that there is a sharp distinction between mathematics and empirical science, we can say that the empirical statistical laws in this derivation are the generalizations about the behavior of standard dice: the probability of 6 is 1/6 and the tosses are independent of one another. The rest is simply arithmetic. The "laws of probability" – such as the multiplication rule – are not empirical laws at all; they are laws of mathematics which have no factual content.

Another example of a D-S explanation would be the derivation of the half-life of uranium-238 from the basic laws of quantum mechanics – which are statistical – and from the height of the potential barrier surrounding the nucleus and the kinetic energies of alpha particles within the nucleus. The answer is approximately 4.5 x 10^9 years. This example is highly theoretical, and consequently distinctly *not* empirical in the narrow sense of "empirical."

As we have seen, the Hempel-Oppenheim article attempted to characterize laws of nature. Because of the very simple formal language employed in that context, there is no possibility of formulating statistical laws within it. The reason, very simply, is that the language does not contain any numerical expressions. Whatever the philosophical merits of using such a language may be, clearly a much richer language is needed to express any important scientific content.[22] Adopting a language that would begin to be adequate for any real science would involve using one in which statistical laws could be formulated. So any language that could contain real universal scientific laws could contain statistical laws as well.

The statistical laws of empirical science are general in the same sense that universal laws are general. The simplest universal law would have the form, "All F are G." It would be formulated using a variable that ranges over all individuals in the universe. The force of the generalization is that, in the entire universe, nothing is an F and not a G or, equivalently, everything is either not an F or else it is a G. The corresponding negative generalization would mean that, in the entire universe, nothing is both F and G. The existential generalization, "Some F are G," would mean that, in the entire universe, at least one thing is an F and also a G. The simplest statistical law could be construed in either of two ways. First, it might be taken to mean, "Every F has a certain propensity to be G," in which case it is strictly analogous to a universal generalization. Second, it might be taken to mean, "A certain proportion of F are G." In that case we can construe it to mean that, among all of the individuals in the universe that are F, a certain proportion are also G. Either way it has the same sort of generality as universal or existential generalizations.[23]

In addition, the same problem about purely qualitative predicates obviously arises in connection with both universal and statistical laws. The same point about not restricting generality by making reference to particular objects, places, or times also applies to statistical as well as to universal laws. And the problem of distinguishing between lawful and accidental statistical generalizations is the same as for universal generalizations. Recalling that the official Hempel-Oppenheim explication of D-N explanation appealed to theories rather than laws, we can say that generalizations involving universal quantifiers, existential quantifiers, mixtures of the two kinds, or probabilities are on a par.

We should also remember that Hempel and Oppenheim confessed their inability to provide an explication for D-N explanation of laws because of the difficulty stated in "the notorious footnote 33." The problem was to make an appropriate distinction between really explaining a law by deducing it from a genuinely more general law and giving a pseudo-explanation by some such device as deducing it from a 'law' that consists of a conjunction of which it is one conjunct. This problem is clearly just as acute for statistical laws as for nonstatistical laws and theories. For reasons of these sorts, I suggested at the outset that we not regard D-S explanations as belonging to a type different from D-N, but treat them rather as a subtype of D-N. The right-hand column of Table 1 can be regarded as representing one kind of explanation, namely, D-N explanation of generalizations. Nowhere in "Aspects" does Hempel offer an answer to the problem stated in the notorious footnote.

2.4.2 The Inductive-Statistical Model

Hempel's main concern in section 3 of "Aspects" is clearly with I-S explanation. As he says, "Ultimately, however, statistical laws are meant to be applied to particular occurrences and to establish explanatory and predictive connections among them" (1965, 381).[24] So let us turn to his explication of that model.

There is a natural way to try to extend the treatment of D-N explanation of particular facts to cover statistical explanation of particular facts. Given the explanation/prediction symmetry thesis, we can say that a D-N explanation of a given fact is a deductive argument showing that the event in question was predictable had the explanatory facts been available early enough. We can then suppose that an I-S explanation is also an argument that would render the explanandum predictable had the explanatory facts been available early enough. In the case of I-S explanation, the explanans must include, essentially, at least one statistical law; as a result, it is impossible to *deduce* the explanandum statement from the explanans. Hempel therefore requires the I-S explanation to be an *inductive* argument that would render the explanandum predictable, not with deductive certainty but with high inductive probability, given the explanans. The simplest kind of example would fit the following schema:

$$\text{I-S} \quad \begin{array}{l} P(G\,|\,F) = r \\ Fb \\ \hline\hline \\ Gb \end{array} \quad [r]$$

The first premise of this argument is a statistical law that asserts that the relative frequency of Gs among Fs is r,[25] where r is fairly close to 1. The double line separating the premises from the conclusion signifies that the argument is inductive rather than deductive. The expression "[r]" next to the double line represents the degree of inductive probability conferred on the conclusion by the premises. Note that I-S explanations are covering law explanations in exactly the same sense as D-N explanations are—namely, each such explanation must contain at least one law in its explanans.

Perhaps a few examples would be helpful. Why, for instance, did Yamamoto suffer severe physical injury in August of 1945? Because he was only a kilometer from the epicenter of the atomic blast in Hiroshima, and any person that distance from the epicenter of an atomic explosion of that magnitude will very probably suffer severe physical injury.

Why, for another example, is the ratio of carbon-14 to other isotopes of carbon in this particular piece of charcoal about half of the ratio of C^{14} to other isotopes in the atmosphere? Because the piece of wood from which this piece of charcoal came was cut about 5730 years ago, and the half-life of C^{14} is 5730 years. The proportion of C^{14} in the atmosphere remains fairly constant, because it is replenished by cosmic radiation at about the same rate as it decays radioactively. While a tree is living it continues absorbing C^{14} from the atmosphere, but when the tree dies or is cut it no longer does so, and the C^{14} content decreases as a result of spontaneous radioactive decay. This example qualifies as I-S rather than D-N for two reasons. First, the explanandum is a particular fact, namely the C^{14} content of one particular piece of charcoal. The fact that many atoms are involved does not make the explanandum a statistical law. Second, the law governing radioactive decay is a statistical law. From this law we can conclude that it is highly probable, but not absolutely certain, that about 1/2 of any reasonably numerous collection of atoms of a given unstable isotope will decay in the period of time constituting the half-life of that particular isotope.

Hempel's main example of I-S explanation is the case of John Jones who recovered quickly from a streptococcus infection. When we ask why we are told that penicillin was administered, and that most (but not all) strep infections clear up quickly when treated with penicillin. If we supply a definite number r for the probability of quick recovery from a strep infection, given that penicillin is administered, this example is easily seen to fit the I-S schema set out above, as follows:

(1) $P(G|F.H) = r$
    Fb.Hb
    $=\!=\!=\!=\!=\!=\!=\!=$ [r]
    Gb

where F stands for having a strep infection, H for administration of penicillin, G for quick recovery, b is John Jones, and r is a number close to 1.

This example can be used to illustrate a basic difficulty with I-S explanation. It is known that certain strains of streptococcus bacilli are resistant to penicillin. If it turns out that John Jones is infected with a penicillin-resistant strain, then the probability of his quick recovery after treatment of penicillin is low. In that case, we could set up the following inductive argument:

(2) $P(G|F.H.J) = r_1$
    Fb.Hb.Jb
    $=\!=\!=\!=\!=\!=\!=\!=$ [r_1]
    Gb

where J stands for the penicillin-resistant character of the strep infection and r_1 is a number close to zero.

This case exemplifies what Hempel calls *the ambiguity of I-S explanation*. We have two inductive arguments; the premises of these arguments are all logically compatible—all of them could be true. The conclusions of these two arguments are identical. Nevertheless, in one argument the conclusion is strongly supported by the premises, whereas in the other the premises strongly undermine the same conclusion—indeed, argument (2) can readily be transformed into an argument that strongly supports the negation of the conclusion of argument (1):

(3) $P(\sim G|F.H.J) = 1 - r_1$
Fb.Hb.Jb

$$\overline{} \quad [1 - r_1]$$

$\sim Gb$

Given that r_1 is close to zero, $1 - r_1$ must be close to 1. Hence, we have two strong inductive arguments with compatible premises whose conclusions contradict one another. This situation is inconceivable in deductive logic. If two valid deductive arguments have incompatible conclusions, the premises of one argument must contradict the premises of the other.

At this point we are confronting one of the most fundamental differences between deductive and inductive logic. Deductive logic has a *weakening principle*, according to which p $\supset$ q entails p.t $\supset$ q, for any arbitrary t whatever. From this it follows that, given a valid deductive argument, it will remain valid if additional premises are inserted, provided none of the original premises is deleted. Probability theory does not have any such weakening principle. Let $P(G|F)$ be as close to 1 as you like, $P(G|F.H)$ may be arbitrarily close to 0.[26] From this it follows that an inductive argument that strongly supports its conclusion may be transformed, by the addition of a new premise consistent with the original premises, into an argument that strongly undermines that conclusion. Before the discovery of Australia, for example, Europeans had strong inductive evidence, based on many observations of swans, for the conclusion "All swans are white." Nevertheless, the addition of one premise reporting the observation of a black swan in Australia not only undermines the inductive conclusion, it deductively refutes it.

Inductive logicians have long recognized this feature of inductive arguments, and have come to terms with it by means of *the requirement of total evidence*. According to this principle, an inductive argument strongly supports its conclusion only if (i) it has true premises, (ii) it has correct inductive form, and (iii) no additional evidence that would change the degree of support is available at the time. Argument (1) above clearly fails to satisfy the requirement of total evidence, for its set of premises can be supplemented as follows:

(4) $P(G|F.H) = r$
       $P(G|F.H.J) = r_1$
       $Fb.Hb$
       Jb
       ================ $[r_1]$
       Gb

where r_1 is very different from r.

The first temptation, with regard to the ambiguity of I-S explanation, might be to impose the requirement of total evidence, and to say that any inductive argument that is to qualify as an explanation must satisfy that requirement.[27] Such a move would be disastrous, for normally when we try to explain some fact we already know that it is a fact. Hence, our body of knowledge includes the conclusion of the argument. If the conclusion is not included among the premises, the requirement of total evidence is violated. If the conclusion is included among the premises, the argument is not inductive; rather, it is a trivially valid deduction — one, incidentally, that cannot even qualify as a D-N explanation, because, if the explanandum is included in the explanans, no law statement can occur *essentially* in the explanans. Consequently, in order to deal with the ambiguity of I-S explanation, Hempel sought a weaker counterpart for the requirement of total evidence that would not rule out altogether the possibility of I-S explanations.

The requirement Hempel saw fit to impose he called *the requirement of maximal specificity (RMS)*. Suppose, referring back to the schema (I-S) given above, that s is the conjunction of all of the premises and k is the body of knowledge at the time in question. "Then," Hempel says, "to be rationally acceptable in [that] knowledge situation . . . the proposed explanation . . . must meet the following condition:

If s.k implies that b belongs to a class F_1, and that F_1 is a subclass of F, then s.k must also imply a statement specifying the statistical probability of G in F_1, say

$P(G|F_1) = r_1$

Here, r_1 must equal r unless the probability statement just cited is simply a theorem of mathematical probability theory. (1965, 400)[28]

The unless-clause in the final sentence is intended to guard against RMS being so strong that it would rule out the possibility of I-S explanation altogether. If we want to explain Gb, then presumably we know that b belongs to the class G, and if we want to use $P(G|F) = r$ as a statistical law, we know that b belongs to F; consequently, k includes the statement that b is a member of $F_1 = F.G$, which is a subclass of F. But, trivially, $P(G|F.G) = 1$; indeed, trivially, all F.Gs are Gs. But this is not an appropriate basis for condemning the original explanation.

Since, Hempel claims, all bona fide I-S explanations must satisfy RMS, and since RMS makes specific reference to a particular knowledge situation, "*the concept of statistical explanation for particular events is essentially relative to a given knowledge situation as represented by a class K of accepted statements*" (1965, 402). Hempel refers to this feature as the *epistemic relativity of statistical explanation*. This relativity has no counterpart in D-N explanation. The reason is that the requirement of maximal specificity is automatically fulfilled in the case of D-N explanation, for, given that all F are G, it follows immediately that all F_1 are G if F_1 is a subclass of F. This is just an application of the weakening principle that was cited above.

We must guard against one easy misunderstanding. Someone might claim that D-N explanations are relativized to knowledge situations because what we take to be a law depends upon what we know at any given time. It is true, of course, that what is considered a law at one time may be rejected at another, and that we can never know for certain whether a given general statement is true. At best, we can hope to have general statements that are highly confirmed and that we are justified in accepting. These considerations apply equally to universal laws, general theories, and statistical laws, and consequently they apply equally to D-N and I-S explanations. The epistemic relativity of I-S explanation refers, however, to something entirely different.

Suppose that we have two putative explanations of two different particular facts, one a D-N explanation, the other I-S. Suppose that each of them has correct logical form—deductive and inductive, respectively. Suppose further that we are prepared, on the basis of our knowledge at the time, to accept the premises of each argument as true. Then, according to Hempel, we are entitled to accept the D-N explanation as correct (recognizing at the same time that we may be mistaken). We are not, however, entitled to accept the I-S explanation as correct, on the grounds just mentioned; in addition, we have to determine whether the statistical law to which we appeal is maximally specific. Whether it is or not depends upon the content of our body of knowledge. A statistical law can be true without being maximally specific. That is why, according to Hempel, we need the requirement of maximal specificity.

With RMS in place, Hempel has provided us with two models of scientific explanation of particular facts, one deductive and one inductive. A comprehensive characterization can be given. Any explanation of a particular occurrence is an argument to the effect that the event-to-be-explained was to be expected by virtue of certain explanatory facts. The explanatory facts must include at least one general law. The essence of scientific explanation can thus be described as *nomic expectability* —that is, expectability on the basis of lawful connections (1962a).

The general conditions of adequacy for scientific explanations set out in the first section of the Hempel-Oppenheim paper can be revised to encompass statistical explanation in the following way:

Logical conditions
1. An explanation is an argument having correct logical form (either deductive or inductive).
2. The explanans must contain, essentially, at least one general law (either universal or statistical).[29]
3. The general law must have empirical content.

Hempel and Oppenheim admit that requirement 3 is vacuous, for it will be automatically satisfied by any putative explanation that satisfies conditions 1 and 2. Any putative explanation that satisfies these conditions qualifies as a potential explanation. In order to qualify as an actual explanation (or, simply, an explanation), a potential explanation must fulfill two more conditions.

Empirical condition:
4. The statements in the explanans must be true.

Relevance condition:
5. The requirement of maximal specificity.

As we have seen, this relevance requirement is automatically satisfied by D-N explanations.

At this juncture, the received view embraced two models of explanation of particular facts, and a large promissory note about D-N explanations of laws (including, of course, statistical laws). The first serious effort to pay off the debt was made by Michael Friedman in 1974 — well into the third decade[30] — and he did not attempt to deal with statistical laws. We shall examine his work in due course, noting for the moment that, even if he succeeds in his effort, the promise of the lower right corner of Table 1 remains unfulfilled.[31]

2.5 Early Objections to the Inductive-Statistical Model

My own particular break with the received view occurred shortly after the incident with Smart that I related at the beginning of this essay. In a paper (W. Salmon 1965) presented at the 1963 meeting of the American Association for the Advancement of Science, Section L, organized by Adolf Grünbaum, I argued that Hempel's I-S model (as formulated in his (1962)), with its high probability requirement and its demand for expectability, is fundamentally mistaken. Hempel's example of John Jones's rapid recovery from his strep infection immediately called to mind such issues as the alleged efficacy of vitamin C in preventing, shortening, or mitigating the severity of common colds,[32] and the alleged efficacy of various types of psychotherapy. I offered the following examples:

(CE-8) John Jones was almost certain to recover from his cold within a week because he took vitamin C, and almost all colds clear up within a week after administration of vitamin C.

(CE-9) John Jones experienced significant remission of his neurotic symptoms because he underwent psychotherapy, and a sizable percentage of people who undergo psychotherapy experience significant remission of neurotic symptoms.

Because almost all colds clear up within a week whether or not the patient takes vitamin C, I suggested, the first example is not a bona fide explanation. Because many sorts of psychological problems have fairly large spontaneous remission rates, I called into question the legitimacy of the explanation proffered in the second example. What is crucial for statistical explanation, I claimed, is not how probable the explanans renders the explanandum, but rather, whether the facts cited in the explanans *make a difference* to the probability of the explanandum.

To test the efficacy of any sort of therapy, physical or psychological, controlled experiments are required. By comparing the outcomes in an experimental group (the members of which receive the treatment in question) with those of a control group (the members of which do not receive that treatment), we procure evidence concerning the effectiveness of the treatment. This determines whether we are justified in claiming explanatory import for the treatment vis-à-vis the remission of the disease. If, for example, the rate of remission of a certain type of neurotic symptom during or shortly after psychotherapy is high, but no higher than the spontaneous remission rate, it would be illegitimate to cite the treatment as the explanation (or even part of the explanation) of the disappearance of that symptom. Moreover, if the rate of remission of a symptom in the presence of psychotherapy is not very high, but is nevertheless significantly higher than the spontaneous remission rate, the therapy can legitimately be offered as at least part of the explanation of the patient's recovery. It follows from these considerations that high probability is neither necessary nor sufficient for bona fide statistical explanation. Statistical relevance, not high probability, I argued, is the key desideratum in statistical explanation.

Henry E. Kyburg, Jr., who commented on my AAAS paper, noticed that a similar point could be made with regard to D-N explanations of particular facts. He illustrated his claim by offering CE-6, the 'explanation' of the dissolving of a sample of table salt on the basis of a 'dissolving spell' (Kyburg 1965). Once the point has been recognized, it is easy to come up with an unlimited supply of similar examples, including my favorite, CE-7, John Jones's 'explanation' of his failure to become pregnant on the basis of his consumption of oral contraceptives.[33]

Scriven's paresis example, CE-5, brought up much earlier in connection with the explanation/prediction symmetry thesis and the notion of expectability, shows that high probability is not required for probabilistic explanation. Scriven's discussion of explanation in evolutionary biology (1959) draws the same conclusion. The appeal to latent untreated syphilis to explain paresis is obviously an appeal to a statistically relevant factor, for the probability that someone with latent untreated syphilis will develop paresis, while not high, is considerably higher than the probability for a randomly selected member of the human population at large. Moreover, it can plausibly be argued that the evolutionary biologist, in explaining occurrences in that domain, also invokes statistically relevant facts. The issue of high probability vs. statistical relevance has thus been joined. It will prove to be a question of considerable importance in the further development of our story.

The Third Decade (1968–77)
Deepening Differences

The third decade is bracketed by Hempel's last two publications on scientific explanation. It opened with his emendation of the requirement of maximal specificity (RMS*), which was designed to fix a couple of technical flaws in his account of I-S explanation (1968).[1] It ended with the publication (in German) of a substantial postscript to section 3 of "Aspects" which is devoted to statistical explanation (1977). Among other things, he retracted the high-probability requirement on I-S explanations.

Insofar as published material was concerned, the issue of high probability vs. statistical relevance remained quite dormant for about five years after I had raised it rather obscurely in (1965).[2] Richard C. Jeffrey (1969) argued elegantly that statistical explanations — with the exception of certain limiting cases — are not arguments, and that the degree of probability conferred upon an explanandum by an explanans is not a measure of the goodness of the explanation. My next publication on the topic was in 1970; it contained a theory of statistical explanation, based upon statistical relevance relations, that was spelled out in considerable detail.[3] An ingenious information-theoretic account of scientific explanation, in which statistical relevance relations play a key role, was published by James G. Greeno in 1970. An account of the statistical-relevance (S-R) model of scientific explanation — based on the three papers by Greeno, Jeffrey, and me — was published as a small book the following year (W. Salmon et al. 1971).

The introduction of the inductive-statistical model of scientific explanation constituted, I believe, a crucial turning point for the received view. Before that, I suspect, many philosophers felt (as I did) quite comfortable with the basic idea of D-N explanation, and confident that an equally satisfactory statistical conception would be forthcoming. As things turned out, the I-S model gave rise to a number of fundamental problems. Three avenues presented themselves as ways of coping with the difficulties. The first was to maintain the received view — to pursue the course already laid out by Hempel — seeking to defend the I-S model against objections, and to repair the faults as they were detected. The second was to attempt to construct an alternative account of statistical explanation of particular facts, such as the S-R model, in the hope of avoiding difficulties encountered by the received view. The third was to reject altogether the possibility of providing probabilistic or statistical explanations of particular facts, thereby maintaining strict deductivism with regard to scientific explanation. The deductivist claims that the box in the lower left-hand corner of table 1 is empty; all legitimate explanations are of the D-N variety (where we persist in regarding D-S explanations as a subset of D-N). These three avenues correspond roughly to three basic conceptions that seem to me, in retrospect, to have dominated the discussion of scientific explanation from the time of Aristotle to the present.[4] Their distinctness stands out clearly, I think, only when we try to give an explicit account of statistical explanation. Statistical explanation is like a Stern-Gerlach magnet that separates the incoming notion of scientific explanation into three divergent beams—the epistemic, modal, and ontic conceptions.

3.1 The Statistical-Relevance Model

If one takes seriously the notion that statistical relevance rather than high probability is the key concept in statistical explanation, one naturally asks what role inductive arguments have in the statistical explanation of particular facts. The simple and rather obvious answer is "None."[5] Approaching Hempel's I-S model from a different angle, in a philosophical gem (1969), Richard Jeffrey argued incisively to that very conclusion. He maintained that when a stochastic mechanism—e.g., tossing of coins or genetic determination of inherited characteristics—produces a variety of outcomes, some more probable and others less probable, we understand those with small probabilities exactly as well as we do those that are highly probable. Our understanding results from an understanding of that mechanism and a recognition of the fact that it is stochastic. Showing that the outcome is highly probable, and that it was to be expected, has nothing to do with the explanation. That the outcome was probable or improbable might be added— in the margin, so to speak—as an interesting gloss on the main text. This paper also provided inspiration for some philosophers who adopted an explicitly mechanistic conception of scientific explanation in the fourth decade.[6]

At about the same time, James G. Greeno (1970) developed an information-theoretic account of statistical explanation in terms of the amount of information transmitted by a statistical theory. Transmitted information is a statistical relevance concept, and among its other virtues, it is defined quantitatively. It can be used to evaluate the explanatory power of statistical hypotheses, but it does not yield quantitative evaluations of individual explanations. Within this approach there is no suggestion that statistical explanations should be identified with inductive arguments.

At around the same time I was trying to work out an alternative to the I-S model that would be based on statistical relevance rather than high probability. The first crucial step is to notice that, whereas high probability involves just one probability value, statistical relevance involves a comparison between two probability values. To construct an explanation based upon statistical relevance we need to compare a posterior probability with a prior probability.

Consider John Jones and his quick recovery from a strep infection. Initially, Jones is simply a person with a strep infection – this is the reference class within which he is included for purposes of assessing the *prior* probability of quick recovery. We may assume, I believe, that this probability is fairly small. This reference class is not, however, homogeneous. We can partition it into two subclasses, namely, those who are treated with penicillin and those who are not. Among those who are treated with penicillin the probability of quick recovery is quite high; among those not so treated the probability is much smaller. Since Jones belongs to the subclass treated with penicillin, the *posterior* probability of his quick recovery is much greater than the prior probability. The original reference class has been relevantly partitioned.

We begin by asking why this member of the class of people with strep infections experienced quick recovery. We answer that he belonged to the subclass who received penicillin, noting that the probability in the subclass is different from the probability in the original reference class. As Hempel points out, however, we are not quite done. The subclass of people who are treated with penicillin can be further relevantly partitioned into those whose infection is of the penicillin-resistant strain, and those whose infection is not. The probability of quick recovery for those who have the penicillin-resistant infection is much smaller than is that for those who have a non-penicillin-resistant infection. We may therefore answer the why-question by stating that Jones recovered quickly from his strep infection because his infection was of the non-penicillin-resistant strain and it was treated with penicillin.

Let us define some of the terms we are using. By a *partition* of a class F we mean a set of subclasses that are mutually exclusive and exhaustive within F – that is, every member of F belongs to one and only one member of the partition. Each of these subclasses is a *cell* of the partition of F. A partition is *relevant* with respect to some attribute G if the probability of G in each cell is different from its probability in each of the other cells. The possibility is left open that the probability of G in one of the cells is equal to the probability of G in the original reference class F.[7] A class F is *homogeneous* with respect to the attribute G if no relevant partition can be made in F. It is *epistemically homogeneous* if we do not know how to make a relevant partition – if, that is, our total body of knowledge does not contain information that would yield a relevant partition. The class is *objectively homogeneous* if it is impossible in principle, regardless of the state of our knowledge, to make a relevant partition. We may then define a *homogeneous relevant partition* as a relevant partition in which each cell is homogeneous. We may then distinguish in an obvious way between *epistemically homogeneous relevant partitions* and *objectively homogeneous relevant partitions*.

Two points should be carefully noted: (1) when an objectively homogeneous relevant partition of a reference class has been given, all relevant factors have been taken into account – i.e., all relevant partitions have been effected – and (2) a relevant partition admits *only* relevant factors, since no two cells in the partition have the same probability for the attribute G.

These concepts deserve illustration. Greeno (1970) offers the example of an American teenager, Albert, who is convicted of stealing a car (D = delinquent offense).[8] Albert is a boy and he lives in an urban environment (San Francisco). Take as the original reference class for purposes of explanation the class of American teenagers (T). If we subdivide it into American teenage boys (M) and American teenage girls (F), we will make a relevant partition, for the probability that a boy will steal a car is greater than the probability that a girl will do so. Moreover, if we partition the class of American teenagers into urban dwellers (U) and rural dwellers (R), we will make another relevant partition, for the delinquency rate in urban areas is higher than it is in rural areas. Taking both of these partitions into account, we have four subclasses of the original reference class – namely, urban male, urban female, rural male, and rural female – and we find that the probability of delinquency is different in these four cells of the combined partition. Symbolically, we can write

$$P(D \mid T.C_i) = p_i; \quad p_i \neq p_j \text{ if } i \neq j$$

where

$$U.M = C_1; \quad U.F = C_2; \quad R.M = C_3; \quad R.F = C_4.$$

This is a relevant partition, but it is obviously not homogeneous, for there are many other factors, such as religion and socio-economic status, that are relevant to delinquency.

Hempel's example of quick recovery (Q) from a strep infection provides a different illustration. In this example, the original reference class is the class of people who have streptococcus infections (S). Clearly, the partition of this class into those who receive treatment with penicillin (T) and those who do not ($\sim$ T) is relevant. Also, a partition can be made into those who have penicillin-resistant infections (R) and those whose infections are not penicillin-resistant ($\sim$ R). If, however, we combine these two partitions to form the cells

$$T.R = C_1; \quad T.\sim R = C_2; \quad \sim T.R = C_3; \quad \sim T.\sim R = C_4$$

the resulting partition is not a relevant partition, for the probabilities

$$P(Q \mid C_i) = p_i$$

are not all different. If the infection is penicillin resistant, it makes no difference (I assume) whether penicillin is administered or not, and (I further assume) the probability of quick recovery is the same for a person with a non-penicillin-resistant infection who does not receive penicillin as it is for anyone with a penicillin-resistant strain. Thus it appears that

$$p_1 = p_3 = p_4 \neq p_2,$$

so our relevant partition of S is

$$S.C_1 = S. \sim R.T; \quad S.C_2 = S.(R.T \vee R. \sim T \vee \sim R. \sim T).$$

This partition is not likely to be homogeneous, for—as Philip Kitcher pointed out—a further partition in terms of allergy to penicillin would be relevant.

The basic structure of an S-R explanation can now be given. Although there are many different ways to ask for a scientific explanation, we can reformulate the request into an explanation-seeking why-question that has the canonical form, "Why does this (member of the class) A have the attribute B?" When an explanation is requested, the request may come in the canonical form—for example, "Why did Jones, who is a member of the class of people who have strep infections, recover quickly?" Or, it may need translation into standard form—for example, "Why did Albert steal a car?" becomes "Why did this American teenager commit a delinquent act?" A is the reference class for the prior probability, and B is the attribute. When a translation of this sort is required, pragmatic considerations determine the choice of an appropriate reference class.

An S-R explanation consists of the prior probability, a homogeneous relevant partition with respect to the attribute in question, the posterior probabilities of the attribute in cells of the partition, and a statement of the location of the individual in question in a particular cell of the partition:

$P(B|A) = p$
$P(B|A.C_1) = p_1; \quad P(B|A.C_2) = p_2; \quad . . .$
$A.C_1, A.C_2, . . .$ constitute a homogeneous relevant partition of A
b is a member of $A.C_k$

Given our definition of a homogeneous relevant partition, we know that our explanation appeals to all of the factors relevant to the explanandum, and only to factors that are relevant.

These relevance considerations enable us to deal not only with such counterexamples to the I-S model as CE-8 (vitamin C and colds) and CE-9 (psychotherapy vs. spontaneous remission), but also with such counterexamples as CE-3 (barometer) that we encountered much earlier in connection with the D-N model. To do so, we need a relation known as *screening off*.

Consider the fact that a sharp drop in the reading on a barometer (B) is highly relevant to the occurrence of a storm in the vicinity (S). The probability of a storm, given the sharply falling barometric reading, is much higher than its probability without reference to the barometric reading, i.e.,

$$P(S|B) > P(S).$$

This is the reason that the barometer is a useful instrument for predicting storms. However, as we remarked above, the reading on the barometer does not explain the storm. The fact is that a certain set of atmospheric conditions, including a sharp decrease in atmospheric pressure, are responsible both for the storm and for the reading on the barometer. Indeed, the actual drop in atmospheric pressure renders the reading on the barometer irrelevant. The probability of a storm (S), given both the drop in atmospheric pressure (A) and the drop in the reading on the barometer (B), is equal to the probability of a storm, given only the drop in atmospheric pressure—i.e.,

$$P(S|A.B) = P(S|A).$$

Thus, although B is relevant to S if A is not taken into account, B is irrelevant to S in the presence of A. This is what it means to say that A screens B off from S. It should be noted that B does not screen A off from S. If the barometer gives a faulty reading because of some malfunction, or if it is placed in a vacuum chamber and thus caused to register a sharp drop, that will have no effect on the probability of a storm. Thus,

$$P(S|A.B) \neq P(S|B).$$

Indeed, destroying all of the barometers inside a thousand mile radius will not affect the probability of storms in the least.

According to the S-R model, no irrelevant factors should be included in an explanation. "To screen off" *means* "to render irrelevant." Accordingly, no factor that is screened off can be invoked legitimately in an S-R explanation. The barometer example is an instance of a large and important class of cases in which a common cause gives rise to two or more effects that are correlated with one another. The atmospheric conditions cause both the barometric reading and the storm. In such cases, we want to block any attempt to explain one of these common effects in terms of others. This can often be accomplished by appealing to the fact that the common cause screens any one of the correlated effects off from any of the others.

Hempel provides another example of just this type. He points out that Koplik spots—small white blisters on the inside surface of the cheek—are a very reliable sign of a measles infection, and they can be used to predict that other symptoms, such as a skin rash or a fever, will soon appear. All of these symptoms, including the Koplik spots, are among the effects of the measles infection. According to the S-R model, then, the Koplik spots have no part in explaining the other symptoms of the disease. Hempel (1965, 374–75) suggests, however, that there might have been occasions, at a time when the nature of the underlying infection was totally unknown, on which it could have been appropriate to invoke them in order to explain the other symptoms.

In my earliest criticism of the I-S model, I suggested that positive relevance rather than high probability is the key explanatory relation. At that time I had not worked out the details of any alternative model. In developing the S-R model—leaning heavily on the work of Jeffrey and Greeno—I came to the conclusion that positive relevance is not required, but that negative relevance could also have explanatory import. The reasons can be given quite straightforwardly and simply. If one makes a relevant partition of a reference class, in some cells the posterior probability of the attribute will be higher than the prior probability, and in some the posterior probability will be lower. This point is obvious in the case of a partition with just two cells. Suppose that we have a large class of coin tosses, and that the probability of heads in that class is ½. Suppose, further, that this reference class can be partitioned into two subclasses, one consisting of tosses of a coin heavily biased toward heads (0.9), the other consisting of tosses of another coin equally biased toward tails. Among the tosses of the second coin some yield heads; the probability for this result in this cell of the partition is 0.1. However, according to an argument of Jeffrey mentioned briefly above, we understand the low-probability outcomes of any given stochastic process just as well as we understand the high-probability outcomes; we understand the heads resulting from tosses of the coin biased for tails just as well as we understand the tails, and by parity of reasoning, we understand the tails resulting from tosses of the coin biased toward heads just as well as we understand the heads. Thus, it appears, we understand all of the results in the original reference class equally well. Some have high probabilities, some have low. Some are found in cells of the relevant partition which reduce their probability, some in cells that increase their probability. Negatively relevant factors as well as positively relevant factors contribute to our understanding of the phenomena we are studying.

We are now in a position to offer a succinct contrast between the I-S and S-R models of statistical explanation.

I-S model: An explanation of a particular fact is an *inductive argument* that confers upon the fact-to-be-explained a *high inductive probability*.

S-R model: An explanation of a particular fact is an assemblage of facts *statistically relevant* to the fact-to-be-explained *regardless of the degree of probability* that results. (W. Salmon et al. 1971, 11)

Both models conform to the covering law conception of scientific explanation, for they both require laws in the explanantia. In particular, in an S-R explanation, the statements of values of prior and posterior probabilities qualify as statistical laws.

3.2 Problems with Maximal Specificity

Hempel's recognition of what he termed "the ambiguity of I-S explanation," which led to the requirement of maximal specificity and the doctrine of essential epistemic relativization of I-S explanation, gave rise to deep difficulties. J. Alberto Coffa was, to the best of my knowledge, the first to recognize their profundity. In a classic article, "Hempel's Ambiguity" (1974) he spelled them out with great care.[9]

Coffa begins by identifying the following schema

$$\begin{array}{ll} \text{(I)} & (x)[Fx \supset Gx] \\ & Fb \\ & \overline{} \\ & Gb \end{array}$$

as the 'basic schema' for the simplest form of D-N explanation (1974, 142). Since the following schema

$$\begin{array}{ll} \text{(II)} & P(G\,|\,F) = r \\ & Fb \\ & \overline{} \;[r] \\ & Gb \end{array}$$

is the obvious inductive analog of schema I, we might be tempted to offer *the naïve model of inductive explanation* in which an I-S explanation is an argument of form II, where r is close to 1, the first premise is a statistical law, and the premises are true. But, as we have seen, given an instance of (II), we can often find another argument of the form

$$\begin{array}{ll} \text{(III)} & P(\sim G\,|\,H) = s \\ & Hb \\ & \overline{} \;[s] \\ & \sim Gb \end{array}$$

where s is close to 1, the first premise is a statistical law, and the premises are true as well. Such pairs of arguments exemplify what Hempel called "the ambiguity of I-S explanation," and its existence utterly undermines the naïve model (Coffa 1974, 143–44).

The existence of such pairs of arguments is an example of what Hempel (1960) called "inductive inconsistency"—namely, two inductive arguments with mutually consistent (indeed, true!) premises that lend strong inductive support to two mutually inconsistent conclusions. Finding this situation intolerable in the domain of scientific explanation, Hempel seeks a ground for rejecting at least one of the two 'explanations.' Having briefly suggested in his first article on statistical explanation (1962) that the requirement of total evidence be invoked, he subsequently formulated (and reformulated (1968)) the *requirement of maximal specificity* to do the job. But, Coffa points out, if the problem were really just the inductive inconsistency, a *much* simpler device would suffice. Since, as Hempel seems to agree, we explain only actual occurrences, not things that do not happen, it is easy to choose the correct explanation from the foregoing pair. It is the one with the true conclusion. Since Hempel did not adopt this obvious and easy solution, Coffa suggests, he must *not* have been exercised about *this* problem. When we look at the way he actually tried to solve the problem, we see that the *real* problem was not the problem of inconsistency, but rather the venerable *reference class problem*.

The problem of choosing the appropriate reference class arose traditionally when adherents of the frequency interpretation of probability, such as John Venn and Hans Reichenbach, attempted to apply probabilities to single cases. What is the probability, for example, that William Smith will still be alive 15 years from now? This is a question of considerable importance to his wife and children, to his employer, and to his insurance company. The problem is that he belongs to many different reference classes, and the frequencies of 15-year survival may differ greatly from one of these classes to another. Within the class of American males it is fairly high, among 40-year-old American males it is even higher, among grossly obese Americans it is not nearly as high, and among heavy cigarette-smoking individuals it is still different. If we are going to try to predict whether he will be alive in 15 years, or estimate his chances of surviving that long, we will want to take into account all available relevant evidence.

Starting with a very broad reference class, such as the class of all Americans, we should partition it in terms of such factors as sex, age, occupation, state of health, etc., until we have taken into account all known factors that are statistically relevant. By the same token, we do not want to partition the class in terms of factors that are known to be irrelevant to his survival, or in terms of factors whose relevance or lack thereof is unknown. We may say that the rule for assigning a probability to a single case is to refer it to the *broadest homogeneous reference class* available, where it is understood that the class is *epistemically* broad and homogeneous—i.e., we have not used any partitions that are not known to be relevant and we do not know how to make any further relevant partitions (see W. Salmon 1967, 90–96, and Salmon et al. 1971, 40–47). Since we are concerned here with prediction, the maxim is to use all available evidence; epistemic relativization is entirely appropriate.

The problem of choosing the correct reference class for explanatory purposes is quite different. When we want to explain some fact, such as John Jones's rapid recovery from his strep infection, we already know that his recovery has occurred—i.e., the explanandum is already part of our total evidence—so we must not appeal to that portion of our total evidence. The reference class problem becomes the problem of determining precisely to what part of our total evidence we are allowed to appeal in constructing an I-S explanation. Hempel's RMS and its successor RMS* were designed to accomplish this aim. Roughly speaking (referring back to the schema II), both versions of the requirement of maximal specificity stipulated that, *in a knowledge situation K*, every known relevant partition of F, except one in terms of G or in terms of properties *logically* related to G, be made. The idea of this requirement is to tell us what factors must be taken into account and what factors must not be taken into account in constructing I-S explanations. Hempel attempted to prove, for each version of the requirement, that it would eliminate the unwholesome ambiguity. RMS* was devised in response to a counterexample to RMS constructed by Richard Grandy (Hempel 1968).

The profound problem Coffa recognized arises, not from the observation that *many* explanations that fit schema II are subject to ambiguity, but from Hempel's assertion that *all* of them are. It is a striking fact that Hempel does not offer arguments to support this transition from many to all (Coffa 1974, 151). Yet, this is the crucial issue, for it is only from the assertion of *universal* ambiguity that the doctrine of *essential* epistemic relativity arises. So Coffa does two things. First, he attempts to reconstruct conjecturally the kind of considerations that may have led Hempel to his conclusion of essential epistemic relativity (151–55), and second, he maintains that it leads to disaster for the theory of I-S explanation (155–60).

Coffa argues that Hempel was looking for some condition to impose on the law-premise of schema II—"$P(G|F) = r$"—that would guarantee that the problem of ambiguity would not arise. RMS (later, RMS*) were the conditions he proposed. Each contains a reference to a knowledge situation, thus implying that the explanation in question would be epistemically relativized. It appears, then, that Hempel felt that no adequate substitute for RMS could exist that did not also contain reference to a knowledge situation. Consequently, according to Hempel, I-S explanation is *essentially epistemically relativized*.

Both Coffa and I agreed that epistemic relativization of probabilistic explanation could be avoided—though we took different tacks in trying to do so—but before looking at them we should consider the consequences of accepting epistemic relativity. In (1971) I thought that nothing of much importance hinged on the issue of epistemic relativity; Coffa's 1974 article changed my mind radically on this point (W. Salmon 1974).

Coffa begins his discussion of epistemic relativization of I-S explanation by analyzing the distinction between epistemic and non-epistemic concepts:

> It is an obvious fact that the meaning of some expressions or concepts can be given without referring to knowledge, whereas that of others cannot. Let me call the latter epistemic and the former non-epistemic. Examples of non-epistemic expressions are easy to find. 'Table,' 'chair,' 'electron,' according to

many, 'truth,' would be typical instances. Examples of epistemic notions are also readily available. The best known instance may be that of the concept of confirmation. Although the syntactic form of expressions like 'hypothesis h is well-confirmed' may mislead us into believing that confirmation is a property of sentences, closer inspection reveals the fact that it is a relation between sentences and knowledge situations and that the concept of confirmation cannot be properly defined (that is, its meaning cannot be given) without a reference to sentences intended to describe a knowledge situation. . . .

Having introduced the distinction between epistemic and non-epistemic concepts, we go on to notice that there is a further interesting distinction to be drawn within the class of epistemic notions based upon the kind of role knowledge plays in them. On the one hand there are those epistemic notions in which knowledge enters essentially as an argument in a confirmation function, or, equivalently, as an ingredient in a statement of rational belief. And then there is the obscure and largely unintelligible remainder.

In the first group we find a significant example provided by Hempel's theory of deductive explanation. After introducing his non-epistemic notion of D-N explanation Hempel went on to say that he could define now the concept of a well-confirmed D-N explanation, a well-confirmed D-N explanation in a tacitly assumed knowledge situation K being, in effect, an argument which in knowledge situation K it is rational to believe is a D-N explanation, i.e., a true D-N explanation. In precisely the same fashion we could correctly and uninterestingly define the concepts of well-confirmed table, well-confirmed chair and well-confirmed electron, given that we started by having the concepts of table, chair and electron. *Since we can only have reason to believe meaningful sentences, a confirmational epistemic predicate is an articulation of independently meaningful components.*

Of course we can understand what a well-confirmed chair is because we began by understanding what a chair is. If 'x is a chair' had not had a meaning, we would not even have been able to make sense of the statement of rational belief made about it. Similarly, we can understand, if not appreciate, the notion of well-confirmed D-N explanation, because we were told first what kind of thing a D-N explanation is.*

Now we are in a position to state Hempel's thesis of the epistemic relativity of inductive explanation. As a consequence of his analysis of the phenomenon of ambiguity, Hempel concludes that the concept of inductive explanation, unlike its deductive counterpart, is epistemic; and he goes on to add that it is not epistemic in the sense in which well-confirmed deductive explanations are.

---

*Prof. Gerald Massey has drawn my attention towards the apparent opacity of epistemically relativized predicates. He has pointed out that this raises serious doubts concerning the possibility of viewing them as expressing properties. (Coffa's note).

The concept of inductive explanation is a non-confirmational epistemic concept. . . .

As Hempel is careful to point out, this means that there is no concept that stands to his epistemically relativized notion of inductive explanation as the concept of true D-N explanation stands to that of well-confirmed D-N explanation. *According to the thesis of epistemic relativity there is no meaningful notion of true inductive explanation.* Hence, we couldn't possibly have reasons to believe that anything is a true inductive explanation. Thus, it would be sheer confusion to see inductive explanations relative to *K* in Hempel's sense as those inductive arguments which in knowledge situation *K* it is rational to believe are inductive explanations.

It is clear that, according to Hempel, *there is a remarkable and surprising disanalogy between deductive and inductive explanations.* When somebody asks us to give an account of deductive explanations, we can do so without referring to anybody's knowledge. If asked, for instance, what sort of thing would it be to explain deductively the present position of a planet, we would refer to descriptions of certain nomic and non-nomic facts but never to anybody else's knowledge. This is a desirable feature in a non-psychologistic account of explanation. Yet, according to Hempel, when we ask what an inductive explanation of the same event would look like, there is no way in which an appropriate answer can be given without talking about knowledge; presumably the knowledge available at the time of the explanation. Even more surprisingly, this reference to knowledge does not play the standard role that such references usually play, to wit, that of providing the epistemic platform for a judgment of rational belief. What role such reference plays is a question which deserves serious attention, since here we find the Achilles' heel of Hempel's whole construction. (1974, 148–50, emphasis added)

Having established the distinction between epistemic and non-epistemic concepts, and having pointed out, within the former category, the distinction between confirmational and non-confirmational concepts, Coffa now announces the result he intends to draw.

I have argued that from a certain problem Hempel felt forced to draw the conclusion that the notion of inductive explanation is epistemically relativized. . . . Now, I would like to explain why I find Hempel's conclusion worth avoiding. I will try to convince you that *to accept Hempel's thesis of epistemic relativity amounts to accepting the claim that there are no inductive explanations*, the concept of I-S explanation relative to *K* functioning as a placebo which can only calm the intellectual anxieties of the uncautious user. If I am right, anyone willing to hold that there are inductive explanations will have to begin by spotting a flaw in Hempel's argument. . . .

Maybe the best way in which I can briefly convey my feelings about the oddity implicit in Hempel's theory of inductive explanation is by noting, that in my view, Hempel's decision to develop a theory of I-S explanation relative to *K* after having argued that the notion of true inductive explanation makes no sense, seems comparable to that of a man who establishes conclusively that Hegel's philosophy is strict nonsense, and then proceeds to devote the rest of his life to produce the definitive edition of Hegel's writings. (1974, 155, emphasis added)

The strategy Coffa adopts to carry out his assault on Hempel's I-S model of explanation and, indeed, any essentially epistemically relativized model of explanation, is to pose a fundamental challenge for Hempel.

Now the question I would like to put to Hempel is the following. Take any I-S explanation relative to *K,* for some given *K*. . . . What is it about this inductive argument that makes it an explanation of its last formula? What reason could anyone have to say that it is an explanation of its conclusion?

It is not difficult to answer this question when we pose it, not for the inductive, but for the deductive case. If one asks, for example, what reason we have to believe that a causal deductive explanation explains its explanandum, the answer is that its premises identify certain features of the world that are nomically responsible for the occurrence of the explanandum event.

Could we say, as in the deductive case, that I-S explanations relative to *K* explain because their premises somehow identify features of the world that are nomically responsible for the explanandum event? Certainly not. This is what we vaguely conceived to be possible while tacitly espousing the naïve model, until Hempel shattered our illusions to pieces by focusing the reference class problem on the theory of explanation. Indeed, *if there is no characterization of true inductive explanation, then it must be because there are no such things that go on in the non-epistemic world of facts that can inductively explain the event.* For if there were such non-epistemic goings on, their characterization would be a characterization of true inductive explanation. Thus, the possibility of a notion of true explanation, inductive or otherwise, is not just a desirable but ultimately dispensable feature of a model of explanation: it is the *sine qua non* of its realistic, non-psychologistic inspiration. It is because certain features of the world can be deterministically responsible for others that we can describe a concept of true deductive explanation by simply describing the form of such features. If there are features of the world which can be non-deterministically responsible for others, then we should be able to define a model of true inductive explanation. And, conversely, if we could define a model of true inductive explanation, there could be features of the world non-deterministically responsible for others. The thesis of epistemic relativity implies that, for Hempel, there are no such features. What, then, is the interest of I-S explanations relative to *K* ? Surely not, as we have seen above, that in

knowledge situation K we have reason to believe that they are inductive explanations. Then what? We detect in Hempel's writings not even a hint as to what
an answer to this question might be. (1974, 157–58, emphasis added)

I have quoted from Coffa at length to convey the subtlety and cogency of his argument, and to capture the style and clarity of his exposition. I think his conclusion
is entirely correct.

Now that we have exhibited the consequences of the doctrine of essential
epistemic relativity of I-S explanation, let us examine more closely the considerations that seem to have motivated that doctrine. Coffa has shown beyond doubt
that it arises somehow from the reference class problem. Looking at the problem
in the context of the frequency interpretation of probability, in which it originally
arose, the problem seems to be that any reference class we choose will be inhomogeneous.

Coffa considers the example of Jones who is both a Texan and a philosopher.
Given that most Texans are millionaires and that most philosophers are not millionaires, we have a naïve inductive explanation of Jones being a millionaire and
of Jones not being a millionaire (1974, 144). Obviously, neither the class of
Texans nor the class of philosophers is homogeneous with respect to being a millionaire, and it appears that there will always be ways of partitioning any given
subclass that will yield different relative frequencies. Perhaps Jones was born at
the moment a Chinese mandarin sneezed, and perhaps the class of Texas philosophers born at any such moment had a proportion of millionaires different from
that in the class of Texan philosophers (1974, 154). Coffa offers a suggestion—he
does not spell it out in detail—that the main difficulty can be circumvented by
thinking of probabilities in terms of propensities rather than frequencies. I shall
return to this idea below. In the meantime, let me follow the frequentist line a
little further.

If we consider some less fanciful cases, such as Greeno's delinquency example
or Hempel's example of the quick recovery from a streptococcus infection, a plausible line of argument emerges. In explaining why Albert stole a car, we saw that
there were many relevant factors that could be summoned—sex, age, place of
residence, socio-economic status of the family, religious background, etc. The
list seems almost endless. Whenever we arrive at a narrower reference class as
a result of another relevant partition, if that class still contains youths who do not
commit delinquent acts, we are tempted to suppose that there is some further factor that relevantly distinguishes Albert from the non-delinquents. The process of
relevant partitioning will end only when we have found a reference class in which
all members are juvenile delinquents. At that point, it may be assumed, we will
have a bona fide explanation of Albert's delinquency. It should be noted emphatically that *the resulting explanation is not inductive, it is deductive-nomological*.
Obviously, whenever we reach a universal generalization of the form "$(x)[Fx \supset$

Gx]" we have a reference class F that is trivially homogeneous with respect to the attribute G, for every F is a G, and so is every member of every subclass of F. For this reason, D-N explanations automatically satisfy RMS. A similar analysis could plausibly be offered in the case of Jones's quick recovery from his strep infection.

One reason for claiming that I-S explanations are essentially epistemically relativized is the supposition that the statistical generalization among the premises does not contain a homogeneous reference class—i.e., that the reference class can, in all cases, in principle, be relevantly partitioned, even if we do not know how to do it at the time. One obvious motivation for such a claim would be a commitment to determinism. According to the determinist, every event that happens—including Albert's theft of an auto and Jones's quick recovery from his strep infection—are completely determined by antecedent causes. Those causes define a reference class in which every member has the attribute in question. If we construct an I-S explanation, the statistical law that occurs as a premise must embody an inhomogeneous reference class—i.e., one that could be relevantly partitioned. The only reason for using such a reference class would be ignorance of the additional factors needed to effect the further partition.

If this were the situation, we could easily understand the epistemic relativity of I-S explanations. If determinism is true, all bona fide explanations are deductive. I-S explanations are not well-confirmed inductive explanations; they are simply incomplete D-N explanations. An I-S explanation is analogous to an enthymeme—a deductive argument with missing premises. By supplying missing premises we can make the argument more complete, but when all of the missing premises are furnished the argument is no longer an enthymeme. At that point it becomes a valid deductive argument. There are no valid enthymemes, for, by definition, they lack premises needed for validity.

Similarly, an I-S explanation is simply an incomplete D-N explanation as long as it embodies an inhomogeneous reference class. As more and more relevant partitions are made, it approaches more closely to a D-N explanation, but it ceases to be an I-S explanation when homogeneity is achieved because, according to the deterministic assumption, homogeneity obtains only when the statistical law has been transformed into a universal law. From the determinist's standpoint, then, there are no genuine inductive explanations; the ones we took to be inductive are simply incomplete deductive explanations. It is easy to see what epistemic relativity amounts to in the context of determinism. When I posed the situation in this way to Hempel—suggesting that he was implicitly committed to determinism (1974)—he informed me emphatically that he is not a determinist (private correspondence).

An indeterminist, it seems to me, is committed to the view that there are reference classes in bona fide statistical laws that are homogeneous—not merely epistemically homogeneous, but objectively homogeneous. One might suppose that the class of carbon-14 nuclei is homogeneous with respect to spontaneous radioactive decay within the next 5730 years (the half-life of C^{14}) – that there is, in principle, no way to make a relevant partition. Unlike William Smith (with whose survival we were concerned), C^{14} atoms do not age, suffer ill health, indulge in hazardous occupations, etc. Their chances of suffering spontaneous decay are remarkably unaffected by their environments. If quantum theory is correct, there is no way to select a subclass of the class of C^{14} nuclei in which the probability of spontaneous decay within the next period of 5730 years is other than one-half. In other words, modern physics strongly suggests that there are nontrivial cases of objectively homogeneous reference classes.

If one accepts the notion that there may be objectively homogeneous reference classes, then one might construct an unrelativized model of I-S explanation along the following lines:

An argument of the form

$$
\begin{array}{l}
\text{(II)} \quad P(G\,|\,F) = r \\
\phantom{\text{(II)} \quad} Fb \\
\phantom{\text{(II)} \quad} \rule{3cm}{0.4pt} \quad [r] \\
\phantom{\text{(II)} \quad} Gb
\end{array}
$$

is an I-S explanation of Gb, provided that (1) the first premise is a statistical law, (2) both premises are true, (3) r is close to 1, and (4) F is an objectively homogeneous reference class with respect to G.

Since this characterization makes no reference to a knowledge situation, it can be taken as an explication of a *true I-S explanation*. An argument of form (II) is a *well-confirmed I-S explanation* with respect to a knowledge situation K if the information in K provides good grounds for believing that it satisfies the conditions following schema (II) above. An argument of form (II) that satisfies conditions (1)–(3), but not condition (4), is – at best – an *incomplete I-S explanation*. An argument of form (II) that qualifies as an incomplete I-S explanation could be considered an *optimal I-S explanation with respect to knowledge situation K* if the reference class F is epistemically homogeneous – that is, in knowledge situation K we do not know how to effect a relevant partition (although a relevant partition is possible in principle). In the foregoing explication, the role of RMS is taken over by condition (4).

As I understood his communication, Hempel implicitly rejected the foregoing strategy, not because of a commitment to determinism, but, rather, because of deep doubts about the intelligibility of the notion of objective homogeneity. On the one hand, these doubts were surely justified as long as no one had produced a reasonably clear explication of the concept. As I learned in the process of attempting to provide such an explication, it is no easy task. I shall return to this problem below. On the other hand, there were two strong reasons for thinking that objective homogeneity is an intelligible concept. First, in the trivial case of universal generalizations (affirmative or negative), the concept is clearly applicable. Second, in a vast range of nontrivial cases, we can easily see that the negation of the concept is applicable. We can be quite certain that the class of American teen-agers is not homogeneous with respect to delinquency, and that the class of humans is not homogeneous with respect to survival for an additional 15 years (W. Salmon 1984, 53–55). It would be surprising to find that the negation of a meaningful concept were unintelligible. So one wonders whether a deterministic prejudice is not operating after all. As Peter Railton (1980) most aptly remarked, some people who do not hold the doctrine of determinism are nevertheless held by it.

It seems to me that proponents of either the I-S model or the S-R model must face the reference class problem. Since the S-R model appealed to objective homogeneity, it was clear that an explication of that concept was sorely needed, and I attempted to offer one (1977).[10] The I-S model (in Hempel's epistemically relativized version) also required a homogeneity condition, but, of course, an epistemic one. RMS was introduced for this purpose. It seems to me, however, that neither RMS nor RMS* was adequate even to Hempel's needs, and that examining its inadequacy provides a good start on the analysis of objective homogeneity. For this purpose, we will need to look at RMS in detail.

When Hempel first elaborated his I-S model of explanation, he realized that a crucial problem concerning relevance arises. He sought to deal with it in terms of RMS. The basic idea behind that requirement is that, subject to certain important restrictions, all available relevant knowledge should be brought to bear when I-S explanations are constructed. The problem concerns these restrictions.

The most obvious consideration, as we have seen, is that when we attempt to explain any given fact we already know that it is a fact—for example, when we try to explain John Jones's rapid recovery from his strep infection (Gb), we know that the rapid recovery has occurred. Thus, if we were to bring to bear all of our relevant knowledge, we would have to place him in the class of people who have strep infections (F), who are treated with penicillin (H), and who experience quick recoveries (G). Relative to that class, the probability of his rapid recovery is trivially equal to 1; indeed,

$$P(G|F.H.G) = 1$$

is a theorem of the mathematical calculus of probability. The fact that he belongs to that class clearly has no explanatory value. Let us recall the formulation of RMS. With respect to arguments that satisfy schema (II), let s be the conjunction of the premises, and let k be a statement that is equivalent to the content of our body of knowledge K. RMS says:

If s.k implies that b belongs to a class F_1, and that F_1 is a subclass of F, then s.k must also imply a statement specifying the statistical probability of G in F_1, say

$$P(G \mid F_1) = r_1$$

Here, r_1 must equal r unless the probability statement just cited is simply a theorem of mathematical probability theory. (Hempel 1965, 400)

If the restriction stated in the unless-clause were not incorporated in RMS, there could be no I-S explanations.

The unless-clause is not, however, strong enough. Suppose, for instance, that John Jones is a prominent public figure whose illnesses are newsworthy. On the evening of his recovery the local television news program reports that fact. Thus, he belongs to the class of individuals who have strep infections (F) treated by penicillin (H) whose quick recoveries are reported by reliable news media (K). Let us assume, at least for the sake of the example, that the probability that a quick recovery took place (G), given that it is so reported (K), is virtually 1; hence, it is greater than the probability of the quick recovery, given only the administration of penicillin. Since it is not a theorem of the probability calculus that $P(G \mid F.H.K) = 1$, Hempel's unless-clause does not block the use of that relation in connection with RMS. Thus, RMS would disqualify Hempel's original example, even though the report of Jones's recovery is irrelevant to the explanation of the recovery. Although the news report may be the source of our knowledge that Jones has recovered quickly, it plays no part in telling us why the quick recovery occurred. Though it may answer an *evidence-seeking* why-question, it contributes nothing toward an answer to the *explanation-seeking* why-question. Thus, the unless-clause needs to be strengthened.

In a probability expression of the form "$P(Y \mid X)$" X is known as the *reference class* and Y as the *attribute class*. The problem RMS is intended to address is that of choosing a statistical law with a suitable—i.e., *maximally specific*—reference class for an I-S explanation of any particular fact. A reference class F is maximally specific with respect to any attribute class G if our available knowledge does not provide any way of making a relevant partition of it—i.e., to pick out a subset of F in which the probability of G is different than it is in F. But we cannot allow complete freedom in the choice of properties in terms of which to make such partitions. Certainly the restriction formulated in Hempel's unless-clause is required, but as our example has just shown, additional restrictions are also needed. Even if a relevant partition of a reference class can be made—but only in terms of information not available until after the occurrence of the fact-to-be-explained—the reference class should not on that account be ruled out as not maximally specific.

A further restriction is needed, as can easily be seen in terms of another example. Suppose we want to know why, in a genetic experiment, a particular plant has a red blossom. In answer, we are told that, in this species, the color of the blossom is determined by a single gene and that red is a recessive characteristic. In addition, both parent plants had red blossoms. Therefore, barring a fairly improbable mutation, the offspring would have red blossoms. This is a legitimate I-S explanation, for given the statistical laws and antecedent conditions, the explanandum is highly probable, but not certain.

This explanation could, however, be challenged on the basis of RMS. Given certain well-known facts about human vision, we know that red is at the opposite end of the visible spectrum from violet. This is *not* a matter of definition.[11] Therefore, it could be said, we know that the blossom is on a plant both of whose parents had red blossoms, and that the blossom on this plant has a color that lies at the opposite end of the visible spectrum from violet. Given this information (which serves to rule out the occurrence of a mutation affecting the color), the probability that the color of the blossom on this plant is red is unity. Therefore, the reference class in the statistical law in the original explanation was not maximally specific and the putative explanation fails to satisfy RMS.

The patent fault in this sort of use of RMS lies in the fact that knowledge that the color of the blossom is at the opposite end of the visible spectrum from violet is tantamount to knowledge that the color is red. In this example—as I intend it to be construed—the only basis of our knowledge that the color is at the opposite end of the visible spectrum from violet is our observation that the blossom is red. We have no way of knowing independently that the color is at the opposite end from violet. Whenever the blossom has a red color, it has a color at the opposite end of the spectrum from violet. Thus, to cite the fact that the color is at the opposite end from violet as an explanatory fact is to explain the red color on the basis of itself.

The best way to block this sort of difficulty is, it seems to me, to insist that the antecedent conditions in an I-S explanation must be temporally antecedent to the explanandum. Hempel's RMS might, therefore, be amended to read as follows:

> If s.k implies that b belongs to a class F_1, and that F_1 is a subclass of F, then s.k must also imply a statement specifying the statistical probability of G in F_1, say
>
> $$P(G|F_1) = r_1$$
>
> Here, r_1 must equal r unless the probability statement just cited is simply a theorem of mathematical probability theory, *or unless b's membership in F_1 cannot be known before its membership in G has been ascertained.* [12]

This version of RMS is obviously relativized to the state of knowledge of the explainer, but Hempel's versions all had that characteristic as well. I am inclined to think that this formulation adequately expresses the intent of Hempel's requirement.[13] It should be recalled, however, that Hempel explicitly rejected a similar temporal constraint on the 'antecedent' conditions in D-N explanations of particular facts.

To make good on the homogeneity requirements for the S-R model of scientific explanation, as remarked above, I attempted to explicate a fully objective concept of homogeneity (1977).[14] The general idea is that, if determinism is true, no explanations of particular facts will be irreducibly statistical; every statistical explanation can, in principle, be supplemented with additional information so as to be transformed into a D-N explanation. If indeterminism is true, some explanations will be irreducibly statistical—that is, they will be full-blooded explanations whose statistical character results not merely from limitations of our knowledge, but rather from the fact that there are no additional factors that would make it possible in principle to beef them up, thereby transforming them into D-N explanations. Whether determinism or indeterminism is true is a question of fact; I believe that the evidence points toward indeterminism, but that is not crucial for this discussion. Whichever is true, we should try to understand clearly the notion of an irreducibly statistical explanation. Speaking roughly and intuitively, given an irreducibly statistical explanation, it must be impossible in principle to identify anything that happens before the event-to-be-explained that would physically necessitate its occurrence.

These general ideas can be stated more precisely in terms of reference classes. Given an explanation that fits form (II), it will be irreducibly statistical if the term F that occurs in the statistical law premise "$P(G|F) = r$" designates a class that cannot be relevantly partitioned, even in principle, with respect to the attribute G. Suppose, that is, that we want to explain why b has the attribute G (at some particular time), and that we have appealed to the fact that b belongs to the class F to do so. If the explanation is to be irreducibly statistical, then prior to the fact that b has the attribute G, b cannot have any attribute in addition to F that has any bearing upon the fact that b has attribute G.

To try to make these ideas more precise, I introduced the concept of a *selection by an associated sequence*. It is, perhaps, best explained in terms of concrete examples. Consider, for instance, all of the basketball games played by the State College team. This class of events can be taken to constitute a reference class F; we are concerned, say, with the attribute G (games won by this team). The members of F (the games) can be taken in chronological order and designated x_1, x_2, . . . [15] Suppose just before each game there is a solo rendition of "The Fight Song" over the public-address system; call this class of events A, and let its members be taken in chronological order and designated y_1, y_2, . . . , where y_i is the rendition of the song immediately preceding game x_i. A is an *associated sequence* for the reference class F, because there is a one-one correspondence (signified by the subscripts) between the members of F and those of A, and each y_i

in A precedes the corresponding x_i in F. Suppose that sometimes the song is sung by a male vocalist and sometimes by a female. We might wonder whether the sex of the vocalist has any bearing upon the chances of winning. So, designating the renditions by male vocalists B, we make a selection S of members of F on the basis of whether the corresponding member of A belongs to B. S consists of exactly those games that are preceded by a male's rendition of "The Fight Song." S is a *selection by an associated sequence*. Formally, x_i belongs to S iff y_i belongs to B. Someone seriously interested in the team's record might wonder whether the sex of the singer has any bearing on the team's success. If $P(G|F.S)$ differs from $P(G|F)$, the sex of the vocalist is relevant to winning. If so, the probability of winning is not invariant with respect to this particular selection by an associated sequence. Given this situation, we must acknowledge that the reference class F is not homogeneous with respect to winning, for we have just discovered a way to partition it relevantly.

Consider Greeno's example of Albert's theft of a car. The reference class F is the class of American teen-agers; let its members be arranged in some sequential order, $x_1, x_2, \ldots$. The attribute G is commission of a car theft. In this case, we should note, the reference class consists not of events but of enduring objects. Nevertheless, the attribute is an event involving a member of the reference class. (The negation of the attribute would be to reach the age of 20 without having stolen a car) We might wonder whether coming from a 'broken home' is relevant to car theft by a teenager. To construct an associated sequence we can again take the class of American teen-agers, but, so to say, at earlier times in their lives. To make the case particularly clear, we might specify that the breakup of the home had to occur before the youth in question turned 13. What is crucial is that the selective attribute (coming from a broken home) occur earlier than the attribute of interest (theft of a car). Examples of this sort are most easily handled by reformulating them to make both the original reference class and the associated sequence consist of events (the breakup of a home, the theft of a car, becoming 20 without having stolen a car, etc.).

As another example, let F be a sequence of tosses of a given coin and let G be the coin landing heads up. We wonder whether the outcome of the preceding toss has a bearing on the outcome of this toss. We form the associated sequence A consisting of tosses of the same coin, where each toss x_i is associated with its immediate predecessor $y_i = x_{i-1}$. B is the attribute of landing heads up. The selection S thus picks out the tosses immediately following a toss resulting in heads. If this is a standard coin being tossed in the standard way, then this selection S will not furnish a relevant partition in F, for $P(G|F) = P(G|F.S)$. In this example, as in the former ones, it is crucial that the event used to make the selection occur before the event consisting of the occurrence or non-occurrence of the attribute of interest.

The general idea is that a reference class F is objectively homogeneous with respect to a given attribute G iff the probability of G in F is invariant with respect to all selections by associated sequences.[16] To rule out the counterexamples offered above in connection with Hempel's RMS, we have to ensure that the selection be made on the basis of events prior to those we are trying to explain. Thus, the case of the red blossoms, we do not even have an associated sequence, for the class F under consideration (flowers) constitutes the sequence each member of which has or lacks 'both' the color red and the color at the opposite end of the visible spectrum from violet, and each member has or lacks 'both' colors at the same time. In the case of Jones's quick recovery from his strep infection, we set up another sequence—TV news reports—but it does not qualify as an associated sequence because its members come after the corresponding members of the original reference class of strep infections. To qualify as an associated sequence, the members must precede the corresponding members of the original reference class. Consequently, although the report of Jones's recovery is statistically relevant to his quick recovery, it does not signify any lack of homogeneity in the original reference class.

The most difficult aspect of the explication of objective homogeneity is to impose suitable restrictions on the properties B that may be used to effect the selection S. Consider the probability that a person who contracts pneumonia will die as a result. There are several types of pneumonia—bacterial, viral, and fungal—and I presume that the probability of death of a victim varies from one type to another. Moreover, an individual may have more than one type at the same time; according to a recent report the combination of a bacterial and a viral infection is particularly lethal. Now, if we start with the general reference class of cases of pneumonia, we should partition in terms of the various types of infection, and, no doubt, in terms of various other factors such as the age and general physical condition of the patient, and the sort of treatment (e.g., penicillin) given. But we must not partition in terms of fatal vs. nonfatal infections. The reason is that very definition of the term "fatal" depends not just on the occurrence of an infection, but also on the occurrence or non-occurrence of a subsequent event, namely, death. Once you see how to do it in one or two examples, it is easy to cook up predicates whose applicability to a given event depends upon something that occurs (or fails to occur) at some later time.

The suggestion I offered for the restriction to be imposed is that B must determine an *objectively codefined class*. Very roughly, such a class is one consisting of events whose membership in the class can, in principle, be determined at the time the event occurs. More precisely, given that y_i is the member of an associated sequence corresponding to x_i, it must be possible in principle to set up a physical detecting apparatus (connected to a computer if necessary) that can deliver the verdict of whether y_i has the property B before it receives any information about the occurrence of x_i or any other event that happens later than x_i. My hope is that the resulting characterization of objective homogeneity is unrelativized either to particular knowledge situations or to particular languages (see W. Salmon 1984, chap. 3, for details). Whether the explication as given is satisfactory or not, it does not appear impossible in principle to define an adequate concept of objective homogeneity.[17]

The utility of the notion of objective homogeneity is not confined, of course, to the S-R model of explanation. If this concept is used in conjunction with Hempel's I-S model, the problem of essential epistemic relativization, as analyzed by Coffa, evaporates, for we can then say what constitutes a true I-S explanation. The fact that almost all of our actual I-S explanations are incomplete would not pose any fundamental conceptual problem. At the end of the third decade, Hempel (1977) acknowledged that there might be some nontrivial cases of objectively homogeneous reference classes. By implication he thereby admitted that the concept of objective homogeneity is not unintelligible.

3.3 Coffa's Dispositional Theory of Inductive Explanation

After articulating his devastating critique of the doctrine of essential epistemic relativization, Coffa offered a different solution to the problem of ambiguity of inductive explanation. According to Coffa, a great deal of the trouble Hempel encountered was a result of his implicit identification of the problem of ambiguity with the reference class problem. As we have seen, the reference class problem arises out of the frequency interpretation of probability when we try to apply probabilities to single cases. The situation could have been significantly improved, Coffa suggests, if Hempel had stuck with the propensity interpretation, which can be construed as a single case physical probability concept. In his doctoral dissertation Coffa (1973, chap. IV) argues that an appeal to the propensity interpretation of probability enables us to develop a theory of inductive explanation that is a straightforward generalization of deductive-nomological explanation, and that avoids both epistemic relativization and the reference class problem. This ingenious approach has, unfortunately, received no attention, for it was never extracted from the dissertation for publication elsewhere.

Coffa begins with a critical examination of the D-N model. Consider a deductive-nomological explanation of the simplest form:

$$(1) \quad (x)[Fx \supset Gx]$$
$$Fa$$
$$\overline{}$$
$$Ga$$

for example – one cited by Carnap and Hempel – in answer to the question, "Why did this iron rod expand?" the explanation that it was heated and that "whenever a body is heated, it expands." This latter statement is, of course, the law required in the explanans — in this case, the law of thermal expansion.[18] One might suppose that this law should read

$$\Delta L = k \cdot L_o \cdot \Delta T$$

where L_o is the original length of the rod, k the coefficient of thermal expansion of the substance in question, and ΔT is the amount by which its temperature was raised. The foregoing explanation could presumably be construed as

(2) $\Delta L = k \cdot L_o \cdot \Delta T$
Rod r increased its temperature by ΔT.
_____
Rod r increased its length by $k \cdot L_o \cdot \Delta T$.

This explanation is not acceptable as it stands, Coffa claims, for the statement of the law of thermal expansion is defective; a correct formulation requires what he calls an "extremal clause." It is not true, in general, that bodies expand when heated. The iron rod will not expand when heated if a sufficient compressing force is present. The correct statement is "*In the absence of other changes relevant to length*, an increase in temperature ΔT is physically sufficient for an increase in length $k \cdot L_o \cdot \Delta T$" (1973, 210). The italicized phrase is the extremal clause (EC). Only when the extremal clause is included do we have a bona fide law, and that is what is needed as the first premise of our explanation. Explanation (2) must be amended as follows:

(3)   If EC then $\Delta L = k \cdot L_o \cdot \Delta T$
EC
Rod r increased its temperature by ΔT.
_____
Rod r increased its length by $k \cdot L_o \cdot \Delta T$.

Because the first premise is a law of nature, change of temperature is *nomically relevant* to change of length. EC says that there is no other factor in the situation that is nomically relevant to the change of length. Assuming that EC is true, and that the initial condition stated in the third premise is true, Coffa claims, we now have a bona fide D-N explanation of the increase in length of the rod. Coffa's major thesis about D-N explanation is that *every law involves an extremal clause*, and consequently, *every D-N explanation must contain, as one of its premises, an assertion that the extremal clause holds on the occasion in question.*

An obvious objection can be raised at this point. If the extremal clause has to be incorporated into the law-statement, does that not make the law vacuous? Hempel and Oppenheim, we should recall, explicitly demanded that the law-statement in a D-N explanation have empirical content. The question is, does the extremal clause in the first premise rob the law of thermal expansion of its empirical content?

Coffa meets this objection squarely, citing an argument that had been given by Hempel to the effect that any law containing an extremal (or ceteris paribus) clause can always be protected from refutation—in all cases in which the law seems to fail—by maintaining that, because the extremal clause is not fulfilled, the law is actually correct. Coffa points out that any law in conditional form, even without the extremal clause, can be protected against refutation—in all cases in which it *seems* to fail—by maintaining that its antecedent was unfulfilled. Thus, any law having a conditional structure can be rendered vacuous by the same technique. He then observes that the pertinent question is not whether laws—with or without extremal clauses—*can* be treated in this way, but whether they *must* be so treated.

The answer, he argues, is negative. Consider, for example, Newton's second law of motion. It states that any body of mass m, upon which the net force is F, experiences an acceleration F/m. When this law makes reference to the *net force* acting on a given body, it is implicitly making a statement about all forces, known or unknown, detected or undetected. Nevertheless, we use Newton's laws of motion nonvacuously for predictive and explanatory purposes. In other words, we sometimes explicitly identify a set of forces acting upon a body and assert (corrigibly, of course) that there are no other forces. This latter assertion is an extremal clause.

Suppose Coffa is right in his claim that all D-N explanations involve extremal clauses. How does this fact apply to nondeductive explanations? It applies quite directly. We can think of universal laws as describing physical dispositions. Massive bodies have a disposition to accelerate when forces are applied; metal wires have a disposition to carry electrical current when a potential difference exists between the two ends. In addition to universal dispositions, the world also seems to contain probabilistic dispositions of varying strengths. A fair coin flipped by an unbiased device has a disposition of strength ½ to land heads up; a tritium atom has a disposition of strength ¾ to decay within a period of 25 years.

If we grant that there are both universal and probabilistic dispositions, then it is natural to think of both deductive and inductive explanation in terms of the following schema (1973, 273):

(4) $P(Fa|Ga) = r$
       Ga
       $EC(a,F,G)$
       _____

       Fa

where "$EC(a,F,G)$" is an extremal clause stating that nothing nomically relevant to a's having property F, other than the fact that a has the property G, is present in this explanatory situation. In the case of inductive explanation, the extremal clause does the job of Hempel's requirement of maximal specificity and of my requirement of objective homogeneity. It makes inductive explanation just as objective and epistemically unrelativized as is deductive explanation. Before accepting Coffa's model I would, of course, insist upon a requirement banning inclusion of irrelevant factors, but that requirement is easily imposed.

Coffa's theory of scientific explanation adheres strongly to the ontic conception. Proponents of this conception can speak in either of two ways about the relationship between explanations and the world. First, one can say that explanations exist in the world. The explanation of some fact is whatever produced it or brought it about. The explanans consists of certain particular facts and lawful relationships. The explanandum is also some fact. This manner of speaking will sound strange to philosophers who have been strongly influenced by the thesis that explanations are arguments, or by the deeply linguistic approaches that regard explanations as speech acts. In nonphilosophical contexts, however, it seems entirely appropriate to say such things as that the gravitational attraction of the moon explains the tides, or the drop in temperature explains the bursting of the pipes. The gravitational attraction and the drop in temperature are out there in the physical world; they are neither linguistic entities (sentences) nor abstract entities (propositions). Second, the advocate of the ontic conception can say that an explanation is something—consisting of sentences or propositions—that reports such facts. It seems to me that either way of putting the ontic conception is acceptable; one can properly say either that the explanandum-fact is explained by the explanans-facts or that the explanans-statements explain the explanandum-statement. Coffa frequently adopts the first of these two manners of speaking; nevertheless, he does identify explanations as arguments, and that would seem to commit him, on those occasions, to the second.

The fact that Coffa often identifies the explanans with what produced or brought about the explanandum strongly suggests that his model is a causal model. He also adopts a covering law conception. The premise "$P(Fa|Ga) = r$" is a law-statement, so there must be a nomic connection between G and F, whether the law is universal ($r = 1$) or statistical ($r < 1$). The law-premise contains a symbol standing for probability, and, as I have mentioned, Coffa construes the probabilities as propensities. Coffa maintains that explanations should appeal only to nomically relevant factors—excluding those that are merely statistically relevant. He suggests, for example, that there may be no nomic relationship whatsoever between the sneeze of a Chinese mandarin at the time of one's birth and one's being a millionaire.

It would be natural, I should think, to restrict the laws admissible in such explanations to causal laws—where we admit probabilistic as well as deterministic causes. Propensities, I would suggest, are best understood as some sort of probabilistic causes. However, Coffa does not impose that restriction. According to Hempel's covering law approach, we can explain why a gas has a certain temperature by specifying the volume, pressure, and number of moles. Such an explanation may tell us nothing at all about how the temperature was brought about; consequently, for Coffa, I believe, it should not count as an explanation.

Several comments must be made regarding Coffa's proposed theory of inductive explanation.[19] In the first place, I have great sympathy with the idea that only nomic relevance should have a place in statistical explanation. In articulating the S-R model, I always maintained that it is a covering law model, all of the probability relations appearing in any such explanation being statistical laws. It is, of course, a difficult matter to distinguish statistical laws from true accidental statistical generalizations — at least as difficult as making the corresponding distinction in the case of universal generalizations, which is, as we have seen, an unsolved problem. But switching from statistical generalizations to propensities does not help in making this crucial distinction.

Second, the propensity interpretation does not escape the problem of the single case; indeed, it faces a problem that is the precise counterpart of the reference class problem. Basic to the single case propensity interpretation is the concept of a *chance set-up*. A chance set-up is a device that produces a set of alternative outcomes, each with a determinate probability. When a fair die is rolled from a dice cup in the standard manner, for example, there are six possible outcomes, and this chance set-up has a propensity of ⅙ to produce each of them on a given roll. A chance set-up can be operated repeatedly to produce a sequence of results. However, when we describe the chance set-up, we have to include relevant features and omit various irrelevant ones to determine what constitutes an additional trial on the same chance set-up. In the example of rolling the die, it is important that the die be physically symmetrical and that a standard cup be used, but it is irrelevant whether the device is operated during the day or during the night, or whether the roll is by a left-handed or right-handed player.

Consider a famous example — due to Laplace. Suppose that a coin is about to be flipped. It is known to be asymmetrical; it has a bias either for heads or for tails, but we do not know which. Laplace claimed, on the basis of his notorious *principle of indifference*, that, because of our ignorance, the probability of heads on this toss is ½. A modern propensity theorist would maintain that this chance set-up has a propensity other than ½; it is either greater than ½ or less than ½, but we do not know what it is. I would insist that either of the foregoing claims might be correct, depending upon how we specify the chance set-up. Suppose this coin has been produced by a machine that operates in the following way. Asymmetrical disks are fed into the machine in a random fashion, and the machine stamps the insignia on these disks. Half of the coins get heads on the favored side and half get tails on the favored side. A coin has been chosen randomly from the pile of coins produced by this machine, and it is about to be flipped. If we describe the operation of the chance set-up as picking a coin and flipping it in the standard manner, we do not know whether the propensity is equal to ½ or unequal to ½. If the continued operation of the chance set-up is to flip the same coin again in

the same manner, then the propensity is different from ½. If the continued operation of the chance set-up is to pick another coin from the output of the machine
and flip it just once, then the propensity of heads is ½. Even for the propensity
interpretation, it is necessary to deal with a problem not essentially different from
the reference class problem. I suggest, then, that switching to the propensity interpretation of probability does nothing to ease Hempel's problem of ambiguity.[20]

Third, as Paul Humphreys pointed out to me in conversation, the "propensity
interpretation of probability" is not an admissible interpretation of the probability
calculus at all (see Humphreys 1985). In fact, Coffa had already hinted at this
point in his dissertation (1973, 272–73, note 32).[21] Bayes's theorem is one of the
theorems of the mathematical calculus, and it provides a way of calculating 'inverse probabilities' from 'direct probabilities.' Imagine a factory that produces
corkscrews. It has two machines, one old and one new, each of which makes a
certain number per day. The output of each machine contains a certain percentage
of defective corkscrews. Without undue strain, we can speak of the relative
propensities of the two machines to produce corkscrews (each produces a certain
proportion of the entire output of the factory), and of their propensities to produce
defective corkscrews. If numerical values are given, we can calculate the propensity of this factory to produce defective corkscrews. So far, so good. Now, suppose an inspector picks one corkscrew from the day's output and finds it defective.
Using Bayes's theorem we can calculate the *probability* that the defective corkscrew was produced by the new machine, but it would hardly be reasonable to
speak of the *propensity* of that corkscrew to have been produced by the new machine. Propensities make sense as direct probabilities (because we think of them
as probabilistic causes, I suspect), but not as inverse probabilities (because the
causal direction is wrong).

Fourth, the concept of probabilistic causality is fraught with difficulties. When
Coffa completed his dissertation in 1973 there were three serious theories of
probabilistic causality in the literature—one by Hans Reichenbach (1956, chap.
IV), another by I. J. Good (1960–61), and the third by Patrick Suppes (1970).
These attempts at explication are far from simple. It is not just a matter of picking
out a suitable subset of probabilities and identifying them as probabilistic causes.
I think that each of the theories suffers from fundamental defects. In (1980) I published a critical study of all three and offered some suggestions of my own. We
shall return to this problem in the fourth decade.

Finally, I am not completely convinced that laws of nature always involve extremal clauses, but I shall not attempt to argue that issue here. It is interesting
to note that in 1982 a workshop on The Limitations of Deductivism was held at
the University of Pittsburgh. Both Hempel and Coffa were participants. Hempel
presented a paper in which he claimed that the use of scientific laws and theories,
in conjunction with initial and boundary conditions, to make predictions cannot
be construed as a strictly deductive affair. The basic reason is the need for provi-

sos — qualifications that strongly resemble Coffa's extremal clauses. The proceedings of this workshop are published in (Grünbaum and Salmon 1988). Whether or not extremal clauses are always involved when universal laws are used, they certainly can be invoked in connection with explanations that employ statistical laws. It is evident that, in such cases, establishing the extremal clause is exactly the same as establishing the objective homogeneity of a reference class. Coffa's appeal to propensities and extremal clauses does not escape the immense difficulties associated with objective homogeneity.

At the midpoint of the third decade Coffa offered profound critiques of Hempel's I-S model and my S-R model. As an alternative he produced an extremely suggestive model of causal explanation that included both deterministic and indeterministic causes. The basic unity of his treatment of deductive and inductive explanation is an appealing aspect of his approach. It certainly contains strong anticipations of developments that occurred in the fourth decade. Nevertheless, enormous problems remained. Chief among them, I believe, is the need for a detailed treatment of the causal concepts to which we must appeal if we are to have any satisfactory causal model of scientific explanation.

Another of the many partisans of propensities in the third decade is James H. Fetzer. Along with Coffa, he deserves mention because of the central place he accords that concept in the theory of scientific explanation. Beginning in 1971, he published a series of papers dealing with the so-called propensity interpretation of probability and its bearing on problems of scientific explanation (Fetzer 1971, 1974, 1974a, 1975, 1976, 1977). However, because the mature version of his work on these issues is contained in his 1971 book, *Scientific Knowledge*, we shall deal with his views in the fourth decade. The fourth decade also includes his collaborative work with Donald Nute on probabilistic causality (Fetzer and Nute 1979).

3.4 Explanation and Evidence

One of the first philosophers to attack the Hempel-Oppenheim thesis of the symmetry of explanation and prediction was Nicholas Rescher (1958). At the very beginning of the second decade, he argued that, because of certain fundamental physical asymmetries in nature, many retrodictive inferences can be made with great certainty, whereas predictions are often much more hazardous. Because of a strong commitment to the inferential conception of scientific explanation he maintained that explanations are arguments. Because of a commitment to a rather robust high-probability requirement he claimed that there is a basic difference between explanatory arguments and predictive arguments — namely, that in explanatory arguments the premises must support the conclusion with certainty or near certainty, whereas in predictive arguments we often have to settle for much lower degrees of probability. For purposes of prediction we need only know that a given occurrence is more probable than not—or, in many cases, merely more probable than any other member of a reasonable set of alternatives. Rescher does not exclude the possibility that there are predictions—such as an astronomical prediction of a solar eclipse—that can be made with high degrees of certainty. His claim is simply that many predictive inferences do not have this character. He stresses his view that this kind of asymmetry is a fact of nature rather than a truth of logic.

Given Hempel and Oppenheim's failure to impose temporal constraints on scientific explanations, they leave open the possibility that the so-called *antecedent conditions* appearing in the explanans may actually occur after the explanandum. Recall our discussion of standard counterexample CE-1 (the eclipse) in §2.3 above. Nevertheless, it seems clear that Hempel-Oppenheim, in enunciating their symmetry thesis, did *not* mean to assert that prediction and retrodiction are equally reliable. They meant rather to say that an argument from *temporally prior* antecedent conditions and laws to a subsequent fact may serve in some contexts as a prediction and in others as an explanation. This is surely the way other philosophers, such as Scheffler, construed the symmetry thesis, and Hempel continued to construe it in that way in his "Aspects" essay. Similarly, I take it, as long as he leaves open the possibility of explanation in terms of subsequent conditions, he must say that there could be cases in which an argument from subsequent 'antecedent' conditions and laws would serve as a retrodiction in some circumstances and as an explanation in others. If explanations may sometimes be given in terms of subsequent conditions, then the symmetry condition would need to be formulated in terms not of prediction in the narrow temporal sense, but, rather, in terms of inference from the observed to the unobserved, whether that inference be predictive or retrodictive.

In an article published in the middle of the second decade, Rescher (1963) further elaborates his views on the relationships among prediction, retrodiction, and explanation. To expound his position, he introduces the notion of a *discrete state system*—i.e., a physical system that may assume different states out of a finite or infinite collection of discrete states. He shows how to construct systems in which both prediction and retrodiction are possible, those in which prediction is possible but retrodiction is impossible, those in which prediction is impossible but retrodiction is possible, and those in which both are impossible. The fact that systems of all of these types can be physically realized—by digital computers, for example—undermines the symmetry thesis as he construes it.[22]

Rescher's 1963 article constitutes one main kernel of his book *Scientific Explanation* (1970), which was published near the beginning of the third decade.[23] Another main kernel comes from an article on the nature of evidence (1958). Because Rescher is strongly committed to the main tenet of the received view—that explanations are arguments—he construes the explanatory relation as an evidentiary relation. Given this view, he continues to maintain that explanations in terms of subsequent conditions are just as legitimate as explanations in terms of temporally antecedent conditions, provided a strong argument can be supplied.

In his 1953 book, we recall, Braithwaite raised certain issues about teleological explanation; he asked, in particular, whether explanations in terms of fulfillment of functions or realization of goals can be considered legitimate. He decided not to preclude them even though such explanations appeal to subsequent conditions. Rescher also raises "the much agitated issue of 'teleology' vs 'mechanism' in the theory of explanation," but he acknowledges that "[f]or purposes of the present discussion, the concept of 'teleology' (and its opposite) will be construed in a somewhat artificial chronologized sense" (Rescher 1970, 66). A teleological explanation in this sense is simply one in which the "antecedent conditions" (in the Hempel-Oppenheim meaning of the term) occur later than the event to be explained. Such explanations need not have anything to do with purposes, ends, or functions. Rescher's entire concern is to argue that putative explanations should not be rejected *solely* on the basis of the temporal relation between the explanatory facts and the fact-to-be-explained. In this regard, he supports Braithwaite's claim, but he does not try to give a theory of explanations that appeal to purposes, ends, or functions. He does not address the difficulties—raised by Hempel— regarding the logical structure of functional explanations.

One of Rescher's central concerns revolves around the nature of evidence. His treatment of this topic is considerably more sophisticated than Hempel's. According to the received view, as usually construed, the premises of a D-N explanation constitute conclusive evidence for its explanandum, and the premises of an I-S explanation provide strong evidence for its explanandum. The latter claim amounts to Hempel's high-probability requirement, and Rescher rejects it. He recognizes the need for some sort of relevance requirement in connection with the concept of evidence, and he embodies it in his theory of evidence. In developing this theory he introduces the notions of evidential presumption, supporting evidence, and confirming evidence (1970, 76–77). Briefly, presumptive evidence renders a given hypothesis more probable than its negation; this is a very weak high-probability concept (1970, 78–79). Supporting evidence is a relevance concept; supporting evidence renders the hypothesis more probable in the presence of that evidence than it was in its absence (1970, 80–83). The concept of confirming evidence incorporates these two notions:

> The statement p is *confirming evidence* for q if (1) p is a presumptive factor for q, and (2) p is supporting evidence for q. Thus confirming evidence must at once render its hypothesis more likely than before *and* more likely than not. This two-pronged concept of confirming evidence perhaps most closely approximates to the idea represented in the common usage of "evidence." (1970, 84)

It is remarkable that—to the best of my knowledge—no one (including Rescher and me) noticed any connection between what he was doing in his evidential approach to probabilistic explanation and what those of us who were concerned with statistical relevance were up to. It may have been due, in part at least, to the fact that his investigation was embedded in a theory of evidence, which is naturally taken to be part of confirmation theory, while the S-R model was developed within the theory of scientific explanation. Such compartmentalized thinking seems, in retrospect, quite shocking.

Rescher reflects another aspect of the received view by maintaining the covering law conception of scientific explanation. As he says, "Scientific explanations are invariably subsumption arguments that cite facts to establish an explanandum as a special case within the scope of lawful generalizations" (1970, 9). This leads him to undertake an extended analysis of the concept of lawfulness (1970, 97–121). He takes it as a point of departure—a point acknowledged by most authors who deal with this issue—that lawful generalizations have counterfactual and modal import. Whereas accidental generalizations state only what happens under actual circumstances, laws state what would happen under circumstances that are merely possible, i.e., under counterfactual circumstances. He then argues that empirical evidence can support only the general claim about actual cases, and, consequently, the assertion of lawfulness regarding a generalization goes beyond the empirical evidence. From this consideration he does not conclude that we have super-empirical *knowledge* of lawfulness; instead, he argues, we *impute* lawfulness on the basis of epistemic considerations. Lawfulness is not something that we discover about generalizations; it is something we supply (1970, 107). Lawfulness is mind-dependent (1970, 113–14; see also Rescher 1969). It reflects the use to which we want to put a generalization in organizing and systematizing our knowledge of the world (1970, 107). Skyrms (1980) and van Fraassen (1980), exactly a decade later, offer similarly pragmatic accounts of lawfulness.

In his discussion of the role of laws in scientific explanation, and the modal import of laws, Rescher asserts,

> A recourse to laws is indispensable for scientific explanation, because this feature of nomic necessity makes it possible for scientific explanations to achieve their task of showing not just *what* is the case, but *why* it is the case. This is achieved by deploying laws to narrow the range of possible alternatives so as to show that the fact to be explained "had to" be as it is, in an appropriate sense of this term. (1970, 13–14)

Inasmuch as this passage comes at the conclusion of a discussion of the role of universal *and probabilistic* laws in explanation, it appears that Rescher wants to attribute some kind of modal import to probabilistic laws. He does not say what it is. As we shall see, D. H. Mellor, writing shortly before the close of the third decade, also advocates a modal interpretation of probabilistic explanation (1976).

I shall explain in greater detail below why I find the notion of *degrees of necessity* quite mystifying.

In spite of his strong adherence to the thesis that explanations are subsumptive arguments, Rescher does suggest, from time to time, that scientific explanation involves an exhibition of the mechanisms by which nature works. He makes this point, for example, in connection with the explanatory power of Darwin's theory (1970, 14–15). This discussion suggests that subsumption under nomic regularities is not really all there is to scientific explanation after all; we need to appeal to the underlying mechanisms as well.

In his discussion of discrete state systems, it will be recalled, Rescher provides a variety of types in which prediction and/or retrodiction, and hence, on his view, explanation is impossible. He makes the profound observation that, in the most fundamental sense, we may nevertheless understand completely what goes on in these systems — we know how they work. The issue arises in the context of statistical explanation:

> The consideration of stochastic systems forces us to the realization that scientific understanding can be present despite an impotence to explain (predict, etc.) *even in principle* certain particular occurrences. As regards explanation, this leads us to recognize the fundamentality of description, being able to deploy laws so as to describe the modes of functioning of natural systems. . . .
>
> The root task of science ought thus to be thought of as a fundamentally *descriptive* one: the search for the laws that delineate the functioning of natural processes. . . . It is thus our grasp of natural laws, be they universal or probabilistic — not our capacity to explain, predict, and so on — that appears to be the basic thing in scientific understanding. For it is undeniable that a knowledge of the pertinent laws goes a long way with endowing us with all that we can possibly ask for in the way of an understanding of "the way in which things work" — even in those cases where the specific desiderata of explanation, prediction, etc., are, in the very nature of things, beyond our reach. (1970, 133–34)

Although it is not Rescher's intention to deny the possibility of probabilistic explanation, the foregoing passage does suggest one line of argument that has been adopted by such current deductivists as John Watkins (1984) and Kitcher (Kitcher & Salmon 1989). They maintain that we can provide deductive explanations of lawful statistical regularities by appealing to fundamental statistical laws, but that we cannot give deductive explanations of particular occurrences. Nevertheless, they can maintain, that does not matter, for our knowledge of the lawful statistical regularities provides understanding of how the world works. We have no need, they argue, for any nondeductive type of explanation.

As I shall explain in discussing the opening of the fourth decade, there are at least three main conceptions of scientific explanation: the *epistemic*, the *modal*, and the *ontic*. Although they have been present since the time of Aristotle, they had not been clearly distinguished. After the advent of models of probabilistic or statistical explanation it became crucial to distinguish them. In the foregoing discussion of Rescher's account we have noticed all three.

3.5 Explanations of Laws

Early in our story we called attention to Hempel and Oppenheim's "notorious footnote 33," in which they pointed to a basic difficulty in connection with explanations of laws. The problem is to distinguish the cases in which we genuinely explain a lawful regularity by deductive subsumption under broader regularities from such nonexplanatory moves as deducing a law from a conjunction of itself with another totally unrelated law. Because they saw no solution to this problem, they refrained from even attempting to provide a D-N account of explanations of laws. To the best of my knowledge this problem was not seriously attacked until more than a quarter of a century later—when Michael Friedman (1974) finally took it on. The delay is especially surprising given the fact that, as Friedman notes, explanations of regularities are more usual than explanations of particular facts in the physical sciences.[24]

The basic theme of Friedman's paper is that science explains various phenomena (where phenomena are general regularities) through unification. Citing the kinetic theory of gases as an example, he says,

This theory explains phenomena involving the behavior of gases, such as the fact that gases approximately obey the Boyle-Charles law, by reference to the behavior of molecules of which gases are composed. For example, we can deduce that any collection of molecules of the sort that gases are, which obeys the laws of mechanics will also approximately obey the Boyle-Charles law. How does this make us understand the behavior of gases? I submit that if this were all the kinetic theory did we would have added nothing to our understanding. We would have simply replaced one brute fact with another. But this is not all the kinetic theory does—it also permits us to derive other phenomena involving the behavior of gases, such as the fact that they obey Graham's law of diffusion and (within certain limits) that they have the specific heat capacities that they do have, from the laws of mechanics. The kinetic theory effects a significant *unification* in what we have to accept. Where we once had three independent brute facts—that gases approximately obey the Boyle-Charles law, that they obey Graham's law, and that they have the specific heat capacities they do have—we now have only one—that molecules obey the laws of mechanics. Furthermore, the kinetic theory also allows us to integrate the behavior of gases with other phenomena, such as the motions of the planets and of falling bodies near the earth. This is because the laws of mechanics also permit us to derive both the fact that planets obey Kepler's laws and the fact that falling bodies obey Galileo's laws. From the fact that *all* bodies obey the laws of mechanics it follows that the planets behave as they do, falling bodies behave as they do, and gases behave as they do. Once again, we have reduced a multiplicity of unexplained, independent phenomena to one. I claim that this is the crucial property of scientific theories we are looking for; this is the essence of scientific explanation—science increases our understanding of the world by reducing the total number of independent phenomena that we have to accept as ultimate or given. A world with fewer independent phenomena is, other things equal, more comprehensible than one with more. (1974, 14–15)

It is evident that, if Friedman is successfully to carry out his program, it will be crucial to provide a method for counting independently acceptable phenomena. By "phenomena" in this context we shall understand general uniformities; they can be represented by lawlike sentences. Since *acceptance* is a central notion, we presume that "at any given time there is a set K of *accepted* lawlike sentences, a set of laws accepted by the scientific community. Furthermore, . . . the set K is deductively closed . . . K contains all lawlike consequences of members of K " (1974, 15). Recalling the Hempel-Oppenheim footnote, we immediately ask whether we can reduce the number of phenomena by, for example, deducing the Boyle-Charles law and Graham's law from the conjunction of these two laws, thereby reducing the number from two to one. Clearly, we do not want this kind of move to count as a reduction. Friedman suggests that an appeal to the notion of *independently acceptable lawlike sentences* will enable us to deal with that problem.

Unfortunately, Friedman does not furnish any precise characterization of the of what it means for one sentence to be acceptable independently of another.

Presumably, it means something like: there are sufficient grounds for accepting one which are not also sufficient grounds for accepting the other. If this is correct, the notion of independent acceptability satisfies the following conditions:

(1) If S $\vdash$ Q, then S is not acceptable independently of Q.
(2) If S is acceptable independently of P and Q $\vdash$ P, then S is acceptable independently of Q.

(assuming that sufficient grounds for accepting S are also sufficient for accepting any consequence of S). (1974, 16–17)

The rationale for condition (1) is evident. By virtue of the entailment, grounds for accepting S simply *are* grounds for accepting Q. However, since S is, in many cases, a stronger statement than Q, grounds for accepting Q may well be insufficient grounds for accepting S.

The rationale for condition (2) is not quite as easy to see. Perhaps it can be made clearer by looking at a trivial reformulation:

(2') Given Q ⊢ P, if S is not acceptable independently of Q then S is not acceptable independently of P.

Suppose that S were not acceptable independently of Q. Then grounds for accepting S would be grounds for accepting Q. Because Q entails P, however, grounds for accepting Q would ipso facto be grounds for accepting P. Hence, grounds for accepting S would be grounds for P, and S would not be acceptable independently of P. To look at the condition a bit more formally, let "E ⊨ X" mean "E constitutes sufficient grounds for accepting X." Then, given Q ⊢ P, we have

$$E \vDash Q \rightarrow E \vDash P$$
$$E \vDash S \rightarrow E \vDash Q$$

from which it follows that

$$E \vDash S \rightarrow E \vDash P.$$

This obviously establishes Friedman's second condition in the form (2').

An indispensable requirement for the implementation of this program is to develop a method for counting phenomena (as represented as lawlike sentences). To get on with the job, Friedman introduces the concept of a *partition* of a sentence S. A partition of S is a set Z of sentences such that Z is logically equivalent to S and each S' in Z is acceptable independently of S. Note that the members of the partition do not have to be mutually exclusive. Using this definition, he defines a *K-atomic sentence S* as one that has no partition (in K). This means that there is no set of sentences {S₁, S₂} in K such that S₁ and S₂ are acceptable independently of S and S₁ & S₂ is logically equivalent to S (1974, 17). If, for example, we took the conjunction of the Boyle-Charles law (S₁) and Graham's law (S₂), it would not be K-atomic, for {S₁, S₂} is a partition of it. The question is whether there can be any K-atomic sentences.

Consider any sentence in K of the form

$$(x)[Fx \supset Gx] \tag{1}$$

which is equivalent to

$$(x)[Fx \cdot (Hx \lor {\sim}Hx) \supset Gx] \tag{2}$$

and to

$$\{(x)[Fx \cdot Hx \supset Gx], (x)[Fx \cdot {\sim}Hx \supset Gx]\} \tag{3}$$

If (3) is a partition of (1), then (1) is not K-atomic. But (3) is a partition of (1) unless at least one member of (3) is not a member of K or at least one member of (3) is not acceptable independently of (1). Because (1) entails each member of (3), each member of (3) must belong to K unless it fails to be lawlike.

In private correspondence I posed the foregoing problem to Philip Kitcher (who has written incisively about Friedman's proposal (1976)) as well as to Friedman. Kitcher's response went something like this: If "H" is a reasonable predicate (signifying something like a natural kind), then " ~ H" is apt not to be a reasonable predicate and the second member of (3) may fail to be lawlike.

I would be inclined to reply to Kitcher as follows: One could argue (along the lines of Hempel and Oppenheim as discussed above) that if the first member of (3) is lawlike, so will be the second. If the first member has unlimited scope so has the second; if the first makes no reference to any particular entity, neither does the second; if the first contains only purely qualitative predicates so does the second; etc.[25] Leaving all that aside, however, it is plausible to suppose that a similar partition could be based upon an exclusive and exhaustive disjunction H_1, H_2, . . . , such that, for each i,

$$(x)[Fx . H_ix \supset Gx] \tag{4}$$

is a lawlike sentence. Since each sentence of the form (4) is a logical consequence of (1), each is a member of K. In that case, (4) provides a partition of (1).[26]

Here is Friedman's reply to my original problem:

The problem you raise . . . is just the *kind* of problem I had in mind in requiring that a partition of a sentence S consist of sentences that are acceptable independently of S. In your case, it would seem in general that our ground for accepting the two conjuncts is just that they follow from 'All F are G.' If, on the other hand, we have grounds for accepting the two conjuncts independently — by testing for the two conjuncts directly, say — then it would seem that 'All F are G' is in no way an explanation of the two, but just a summary of what we already know.

I am not satisfied that Friedman's answer is adequate. Consider the generalization (in K), "All humans are mortal." Is it K-atomic? K contains "All men (i.e., human males) are mortal" and "All women (i.e., human females) are mortal." Suppose, as is plausible, that ancient Greek men were male chauvinist pigs. They noticed that each man seemed to die, so by simple induction they concluded that all men are mortal. At this point they could accept "All men are mortal" independently of "All humans are mortal." Some male then made the further generalization that all humans are mortal and it was accepted by the scientific community. Now, I take it, "All men are mortal" is still acceptable independently of "All humans are mortal," but "All women are mortal" is not, for the only basis for asserting this latter statement is that it follows from "All humans are mortal." Thus, {"All men are mortal"; "All women are mortal"} does not constitute a partition of "All humans are mortal," since the second member is not acceptable independently of it. Perhaps this general sentence, "All humans are mortal," is K-atomic.[27]

Suppose that, without changing the membership of the class K in any way, one or more of these MCP Greeks starts noticing that women also die sooner or later, thus accumulating inductive evidence sufficient to accept "All women are mortal" on that basis. Now it is possible to accept that sentence independently of "All humans are mortal," with the result that this latter sentence is not K-atomic, since it now has a partition.

Judging from remarks earlier in his essay, Friedman (1974, 14) — as well as Hempel and Oppenheim — is eager to develop an objective account of scientific explanation. In view of this consideration Friedman would, I should think, regard the order in which the different sorts of evidence for a sentence are actually acquired as irrelevant to the K-atomicity of that sentence. The foregoing example seems to violate this kind of objectivity.

Consider a more serious (quasi-historical) example. Newton's law of universal gravitation states that between any pair of material objects there exists a certain sort of attractive force. Is this a K-atomic sentence? By the time his *Principia* was published, there was considerable evidence to support this law in the domain of pairs of material bodies in which each member of the pair is a large astronomical body (earth, moon, planet, sun, etc.) and in the domain in which one member of the pair is a large astronomical body (earth) and the other a much smaller object (apple, stone, etc.). However, the only evidence for the claim that it also holds in the domain of pairs of which both members are small is the fact that it follows from the general law. Hence, the foregoing tripartite division of the set of all pairs of material bodies {both large, one large and one small, both small} could not have been used at the end of the seventeenth century to provide a K-partition of Newton's law of universal gravitation. However, after the Cavendish torsion balance experiment was performed in the following century there was independent evidence for Newton's law in the third domain of pairs. We could apparently then use as a K-partition: {"Given any two extremely massive material bodies, there exists an attractive force between them . . . "; "Given any two bodies one of which is extremely massive and the other is not, there exists a force . . . "; "Given any two material bodies, neither of which is extremely massive, there exists a force . . . "}. The result is that Newton's law is demonstrably not K-atomic.[28]

If a sentence that was K-atomic (with respect to a given class K of lawlike sentences) becomes non-K-atomic (with respect to the same class), that fact has a profound bearing upon its explanatory import. As we shall see, Friedman offered two definitions of scientific explanation. As we shall also see, Kitcher proved that, under Friedman's first definition of explanation, only K-atomic sentences explain.[29] Consequently, we find a strong presumption that the performance of the Cavendish experiment (along with others, perhaps) robs Newton's law of gravitation of its explanatory import. My personal intuition is that the explanatory power of Newton's law was enhanced as a result of such experiments.

Be that as it may, I must return to my original question, namely, on Friedman's characterization, can there be any K-atomic sentences. I have suggested that among simple generalizations of the form "All As are Bs" there cannot be any. I have not tried to extend the argument to lawlike generalizations of all sorts, but I should think that, if there are any K-atomic sentences, some of them would have this simple form. *If there are no K-atomic sentences, Friedman's program of characterizing scientific explanation in terms of the reduction of the number of independently accepted laws cannot get off the ground.*

It strikes me that Friedman's resolution of the problem is excessively pragmatic for his purposes.[30] If *there exists* evidence sufficient for the acceptance of a generalization having form (1) above, then *there exists* evidence for the independent acceptance of both of the generalizations in the partition (3) above. Which pieces of evidence are found first and which lawlike generalizations are formulated first should be irrelevant if our aim is to formulate an *objective* theory of scientific explanation.

While the difficulties in characterizing K-atomic statements seem serious, I do not mean to suggest that they are insuperable. Let us therefore try to see what Friedman has accomplished if the concept of K-atomic sentence turns out to be viable. The answer, I believe, is that he has solved the problem originally stated in the famous Hempel-Oppenheim footnote. Without going into all of the technical details of his explication, I shall attempt to sketch the development.

Given the notion of a K-atomic sentence, the *K-partition* of a set of sentences Y can be defined as a set Z of K-atomic sentences such that $Y \leftrightarrow Z$. In general there may be more than one K-partition of a given set Y. The *K-cardinality* of a class Y can be taken as the number of members of the smallest K-partition of Y (i.e., the greatest lower bound of the cardinality of K-partitions of Y). A sentence S *reduces* a class Y if the K-cardinality of the union of {S} with Y is smaller than the K-cardinality of Y. Friedman then proceeds to define explanation:

> How can we define *explanation* in terms of these ideas? If S is a candidate for explaining some S' in K, we want to know whether S permits a reduction in the number of independent sentences. I think that the relevant set we want S to reduce is the set of *independently acceptable* consequences of S ($con_K(S)$). For instance, Newton's laws are a good candidate for explaining Boyle's law, say, because Newton's laws reduce the set of their independently acceptable consequences—the set containing Boyle's law, Graham's law, etc. On the other hand, the conjunction of Boyle's law and Graham's law is not a good candidate, since it does not reduce the set of its independently acceptable consequences. This suggests the following definition of explanation between laws:

> (D1) S_1 explains S_2 iff $S_2 \in con_K(S_1)$ and S_1 reduces $con_K(S_1)$..

(1974, 17)

Friedman is not satisfied with this definition, for he finds it too restrictive.

> Actually this definition seems to me to be too strong; for if S_1 explains S_2 and S_3 is some independently acceptable law, then S_1 & S_3 will not explain S_2 — since S_1 & S_3 will not reduce $con_K(S_1$ & $S_3)$. This seems undesirable — why should the conjunction of a completely irrelevant law to a good explanation destroy its explanatory power? So I will weaken (D1) to
>
> > (D1′) S_1 explains S_2 iff there exists a partition Z of S_1 and an $S_i \in Z$ such that $S_2 \in con_K(S_i)$ and S_i reduces $con_K(S_i)$.
>
> (1974, 17)

As Kitcher points out (1976, 211), (D1′) is not just a liberalization of (D1). Having demonstrated that, according to (D1), *only* K-atomic sentences can explain, he points to the obvious fact that, according to (D1′), *no* K-atomic sentences can explain. (D1) and (D1′) are simply different sorts of definition. Kitcher suggests that Friedman's intent might best be captured by a definition whose definiendum consisted of the disjunction of the right-hand sides of (D1) and (D1′). My response to this situation is different. Having already canvassed the many difficulties that arise if we allow explanations to contain irrelevancies, I believe (D1′) should simply be rejected.

If we continue to assume that the problem of K-atomic sentences can be solved or circumvented, then Friedman's theory provides a way for the received view to fill the upper right-hand sector of Table 1 — the D-N explanation of universal laws. It is not clear whether Friedman's approach could be extended to handle the lower right-hand sector — what Hempel called D-S explanation. Since I have maintained above that D-S explanation should be regarded as a species of D-N explanation, the hope would be that Friedman's approach could handle this sector as well. As the end of the third decade approached, the received view had theories of D-N explanation of particular facts, I-S explanation of particular facts, D-N explanation of universal generalizations, and hopes for a theory of D-N explanation of statistical generalizations. Although each of these models suffered from severe difficulties, the received view was at least approaching a complete articulation.

Assuming that Friedman's account of D-N explanation of laws could handle the problem of K-atomic sentences, it appears to be just what Hempel and Oppenheim were looking for in answer to the problem stated in their "notorious footnote 33." However, as Kitcher showed in his critique of Friedman's paper, that theory of explanation of laws is not what we should have wanted, for it is subject to some severe shortcomings.

Kitcher claims that Friedman's definition (D1) is vulnerable to two sorts of counterexamples.

> Counterexamples of the first type occur when we have independently acceptable laws which (intuitively) belong to the same theory and which can be put

together in genuine explanations. The explanantia that result are not K-atomic and hence fail to meet the necessary condition derived from Friedman's theory.

Consider, for example, the usual derivation of the law of adiabatic expansion of an ideal gas, given in books on classical thermodynamics. The explanans here is the conjunction of the Boyle-Charles law and the first law of thermodynamics. These laws are acceptable on the basis of quite independent tests, so their conjunction is not K-atomic. However, the derivation of the law of adiabatic expansion from the conjunction is, intuitively, a genuine explanation. (1976, 209–10)

Kitcher's second type of counterexample concerns the explanation of complex phenomena where, typically, many different laws or theories are brought to bear. He cites, as one instance, the explanation of why lightning flashes are followed by thunderclaps. "The explanation utilizes laws of electricity, thermodynamics, and acoustics, which are independently acceptable."[31]

Kitcher goes on to show—convincingly, in my opinion—that an appeal to Friedman's other definition (D1′) does not help in dealing with the foregoing sorts of counterexamples. He concludes, therefore, that Friedman's account of explanation by unification is unsatisfactory. Nevertheless, Kitcher is firmly committed to the general thesis (with which I am strongly inclined to agree) that unification is a fundamental goal of explanation—that unification yields genuine scientific understanding. In the fifth decade Kitcher articulates his version of the unification theory in Kitcher and Salmon (1989).

3.6 Are Explanations Arguments?

Early in the third decade Jeffrey (1969) argued persuasively that, in many cases at least, statistical explanations are not arguments. His article as well as his thesis was incorporated into the publication in which the S-R model was first set forth in detail (W. Salmon et al. 1971). By the end of the third decade (1977c) I had labeled the claim that scientific explanations are arguments a *third dogma of empiricism* and urged its wholesale rejection.[32] Since the doctrine that scientific explanations are arguments—deductive or inductive—is absolutely central to the received view, my intent was to undermine the orthodox position as deeply as possible. In my zeal to rebut the claim that all explanations are arguments, I argued that no explanations are arguments. This view now seems too extreme; as it seems to me now, some are and some are not.[33] The challenge took the form of the following three questions I hoped would prove embarrassing to devotees of the received view. What I failed to notice until recently is that, while these questions *are* acutely embarrassing with regard to explanations of particular facts, they are innocuous for explanations of laws.

QUESTION 1. Why are irrelevancies harmless to arguments but fatal to explanations?

In deductive logic, irrelevant premises are pointless, but they have no effect whatever on the validity of the argument. Even in the Anderson-Belnap relevance logic, p & q ⊢ p is a valid schema. Consider the following variation on a time-honored argument:

All men are mortal.
Socrates is a man.
Xantippe is a woman.

Socrates is mortal.

It is strange, somewhat inelegant, and possibly, mildly amusing, but it is obviously valid. In contrast, as we have seen, the appearance of an irrelevancy in a D-N explanation can be disastrous.

> The rooster who explains the rising of the sun on the basis of his regular crowing is guilty of more than a minor logical inelegancy. So also is the person who explains the dissolving of a piece of sugar by citing the fact that the liquid in which it dissolved is *holy* water. So also is the man who explains his failure to become pregnant by noting that he has faithfully consumed birth control pills. (1977c, 150)

The situation is no different when we consider inductive relations. There is a *requirement of total evidence* that requires inductive inferences to contain all relevant premises, but it does not ensure against the inclusion of irrelevant premises. When a detective attempting to establish the identity of a murderer comes across evidence that may or may not be relevant—i.e., about whose relevance he or she cannot as yet be sure—it behooves him or her to include it. If it is irrelevant it can do no harm; an irrelevant premise simply does not change the probability of the conclusion. If it turns out to be relevant it may be very helpful.

Where inductive explanation is concerned, again the situation is entirely different. If psychotherapy is irrelevant to the remission of neurotic symptoms, it should not be invoked to explain a patient's psychological improvement. If massive doses of vitamin C are irrelevant to the rapidity of recovery from a common cold, such medication has no legitimate place in the explanation of a quick recovery.

> Inference, whether inductive or deductive, demands a requirement of total evidence—a requirement that *all* relevant evidence be mentioned in the premises. This requirement, which has substantive importance for inductive inferences, is automatically satisfied for deductive inferences. Explanation, in contrast, seems to demand a further requirement—namely, that *only* consider-

ations relevant to the explanandum be contained in the explanans. This, it seems to me, constitutes a deep difference between explanations and arguments. (1977c, 151)

The second query comes in two forms that are so closely related as to constitute one question:

QUESTION 2:   Can events whose probabilities are low be explained?

QUESTION 2': Is genuine scientific explanation possible if indeterminism is true?

The close relationship between these two questions rests upon a principle of symmetry that we have discussed above. Suppose, in a genuinely indeterministic situation, there are two possible outcomes, one highly probable, the other quite improbable. Jeffrey argued, and I fully agree, that in such circumstances, when we understand that both are results of the same stochastic process, we understand the improbable outcome (when it occurs) just as well as we understand the probable outcome (when it occurs). But if probabilistic explanations are inductive arguments, an undesirable asymmetry arises. The explanans confers upon the probable explanandum a high inductive probability, so the explanation is a strong inductive argument. In the case of the improbable outcome, the explanans provides a strong inductive argument for the *non-occurrence* of the event to be explained. The fact that some events have low probabilities seems to cast doubt upon all statistical explanations of individual events if explanations are viewed as being, in essence, arguments.

QUESTION 3:   Why should requirements of temporal asymmetry be imposed upon explanations (while arguments are not subject to the same constraints)?

Among the earliest counterexamples brought against the D-N model of Hempel and Oppenheim was (CE-1), the case of the eclipse. We noted that the occurrence of a given eclipse could be deductively inferred, on the basis of the same laws, either from conditions prior to the eclipse or from conditions subsequent to the eclipse. Only one of these inferences—that from prior conditions—might possibly count as an explanation. A similar point can be made in terms of (CE-2), the flagpole example. Given the angle of elevation of the sun in the sky, the length of the shadow can be inferred from the height of the flagpole, or the height of the flagpole can be inferred from the length of the shadow. The height of the flagpole explains the length of the shadow because the interaction between the sunlight and the flagpole occurs before the interaction between the sunlight and the ground.[34] The length of the shadow does not explain the height of the flagpole because the temporal relation is wrong.

As Rescher had already pointed out, it would be quite mistaken to suppose that inference is in general temporally symmetrical. In a wide variety of cases it is possible to make quite reliable inferences from subsequent facts to earlier occurrences. It is often difficult to predict whether precipitation will occur at a particular place and time; it is easy to look at the newspaper to determine whether rain or snow occurred there yesterday. Dendrochronologists can infer with considerable precision and reliability from the examination of tree rings the relative annual rainfall in certain areas for thousands of years in the past. No one can predict the relative annual rainfall with any reliability for even a decade into the future. The situation is obvious. We can have records—human or natural—of things that have happened in the past. We do not have records of the future, or any comparable resource for making predictions.

We see, then, that explanation involves a temporal asymmetry, and that inference in many cases also involves a temporal asymmetry. What is striking is that the two asymmetries are counterdirected. This would be very strange indeed if explanations are, in essence, arguments. The reason for the contrast is easy to fathom. We explain effects in terms of their causes, but in many cases we infer causes from their effects. Recognizing the crucial role of causal considerations in relation to the third question, I declared that the time had come to put "cause" back into "because" (1977c, 160).

Proponents of the received view had various possible answers to these questions. In response to the first question they could simply impose some sort of requirement that would block irrelevancies. No one (to the best of my knowledge) ever claimed that *all* logically correct arguments with true premises are explanations. As we saw, Hempel and Oppenheim worked hard to characterize the kinds of valid deductive arguments that can qualify as explanations. Advocates of the received view could simply admit that a further requirement, excluding irrelevancies, should be added to the D-N model.[35] Similarly, there are various ways of dealing with irrelevancies in the theory of I-S explanations. A simple way, as I remarked above, would be to amend the requirement of maximal specificity to make it into the requirement of *the maximal class of maximal specificity*.[36]

Several sorts of response to the second question are available to champions of the received view. In the first place, one could simply deny the symmetry principle offered by Jeffrey et al., insisting that we can explain highly probable occurrences, but not those with smaller probabilities. D. H. Mellor (1976) has defended this position. I must reiterate my strong intuition that this response perpetuates a highly arbitrary feature of Hempel's I-S model.

In the second place, one could deny that there are such things as probabilistic explanations of particular facts, insisting that all legitimate explanations are of the D-N variety (perhaps including the D-S type). For reasons that appear to be somewhat different from one another, Wolfgang Stegmüller (1973) and G. H. von Wright (1971) take this tack. I worry about whether this position reflects an anachronistic hankering after determinism, but I shall discuss this thesis in detail in connection with developments occurring in the fourth decade, especially §4.9.

It may turn out, in the end, to be the most promising way to avoid a host of problems associated with statistical explanation.

In the third place, Hempel (private correspondence) suggested the possibility of getting around the problem posed in the second question by reconstruing the concept of an inductive argument. During the late 1940s Carnap developed a detailed and rigorous system of inductive logic that was published in his monumental work *Logical Foundations of Probability* (1950, 2nd ed., 1962). Hempel was strongly influenced by Carnap's theory. One of the peculiarities of Carnap's system is that it has no place for *rules of acceptance* in inductive logic. In deductive logic we think of an argument as a group of statements consisting of a set of premises and a conclusion. If the argument is valid, then, if we accept the premises as true (or well-founded) we should be willing to accept the conclusion as well. It is natural to think of inductive arguments in an analogous way. Under this conception, an inductive argument would be a set of statements consisting of premises and a conclusion. If the argument has a correct inductive form, then, if we accept the premises as true (or well-founded) *and if they comprise all available evidence relevant to the conclusion* we should be willing to accept the conclusion as well. The schemas Hempel has offered in his various discussions of D-N and I-S explanation strongly suggest just such an analogy. In Carnap's inductive logic no such analogy holds. His denial of rules of acceptance for inductive logic means that *there are no inductive arguments* in this straightforward sense. Inductive logic furnishes only degree of confirmation statements of the form "$c(h,e) = r$" where the hypothesis h may be any statement, evidence e any consistent statement, and r any real number between 0 and 1 inclusive. Inductive logic tells us how to compute r. If r has been calculated correctly, it is the degree of confirmation of hypothesis h on evidence e. If e comprises all available evidence relevant to h, then r may be interpreted as a fair betting quotient; that is, $r:1 - r$ constitute fair odds for a bet on the truth of h. But no matter how close r is to 1, we are not permitted to extract the hypothesis h from the degree of confirmation statement and assert it separately as a statement we accept. There are no arguments of the form

$c(h,e) = r$

e

e contains all relevant evidence

$$\overline{\qquad\qquad\qquad\qquad\qquad} \quad [r]$$

h

in Carnap's inductive logic.[37]

If one adopts this sort of inductive logic, it is not altogether clear how there could be any such thing as I-S explanation. For one thing, we could not literally construe I-S explanations as inductive arguments. For another, we would have to abandon the covering law conception, for we could never have any *accepted* law statements to include in the explanans of *any* explanation, D-N or I-S. But one benefit would accrue. There would no longer be any need to impose the high-probability requirement. In Carnap's inductive logic high probabilities do not enjoy special virtues. If inductive logic yields only betting quotients, determining fair odds for wagering, what counts is having a *correct* value, not a *high* value. It is not necessarily unreasonable to bet on improbable outcomes, provided the odds are right. Thus, by adopting this special Carnapian construal of "inductive argument" Hempel can drop the high-probability requirement and avoid the asymmetry between explanations of high-probability and low-probability occurrences. But the price, with respect to the received view of scientific explanation, is rather large (see W. Salmon 1977a).

Question 3, which ties in directly with the role of causality in scientific explanation, raises the most profound issues, I think. The notion that scientific explanation and causality are intimately connected is not new; it goes back at least to Aristotle. As we have seen, some of the earliest objections to Hempel's approach, voiced by Scriven, criticized the received view for its virtual neglect of causality. A little later, when the S-R model was being developed, it seemed clear to me that any adequate theory of scientific explanation must accord to causality a central role. The fact that we were dealing with statistical explanation, in contexts that might in some cases be indeterministic, did not preclude causality. After reading the nearly complete manuscript of Reichenbach's *The Direction of Time* (1956) in the summer of 1952 I was convinced that probabilistic causality is a viable notion. I hoped that such probabilistic concepts as his screening off and conjunctive forks would be helpful in explicating probabilistic causality, but none of it was really worked out at that time (W. Salmon et al. 1971, 76, 81).

Two major factors forced me to focus more carefully upon causality. The first was a brief article by Hugh Lehman (1972) that nicely pointed out the limitations, with respect to scientific explanation, of statistical relevance relations alone. Showing how distinct causal factors could give rise to identical statistical relevance relations, his argument strongly suggested the necessity of appealing to causal mechanisms.

The second factor involved theoretical explanation. Everyone recognizes the obvious fact that among our most impressive scientific explanations are many that appeal to theories that apparently make reference to such unobservable entities as molecules, atoms, and subatomic particles. There is no obvious way in which statistical relevance relations by themselves can account for this feature of scientific explanation. Greeno made a stab at doing something along that line, but his attempt just did not work (see Greeno 1971 and Salmon 1971). I decided to have a try at analyzing theoretical explanation in a paper (whose ambitious title was chosen before the paper was written) for the Conference on Explanation at the University of Bristol in 1973 (W. Salmon 1975). In attempting to cope with that task I found myself deeply involved with causal connections and causal mechanisms. The paper fell far short of its intended goal. But it reinforced in my mind the need to come to grips in a serious way with causal concepts. Theoretical explanation had to wait for the fourth decade (W. Salmon 1978).

3.7 The Challenge of Causality

With causality demanding attention, it was impossible to ignore the issues Hume had raised regarding the nature of causal relations. If we are to succeed in putting "cause" back into "because," we need to understand causal connections. In a great many cases, in everyday life and in science, we take one event to be a cause of another even though they are not spatio-temporally contiguous. In such cases we expect to be able to find some physical connection between them. We flick a switch on the wall by the door and a light on the ceiling at the center of the room lights up. They are connected by an electrical circuit, and a current flows when the switch is closed. Ammonia is spilled in the laundry room and its odor is detected in the hallway. Ammonia molecules diffuse through the air from one place to another. As long as we steer clear of quantum mechanical phenomena, the notion of continuous physical connections between causes and effects seems assured. Hume, in his characterization of cause-effect relations, mentioned priority, contiguity, and constant conjunction. He failed to find any additional feature of the situation that constitutes a causal connection. If the cause and effect are not contiguous, the natural move is to interpolate intermediate causes so as to form a causal chain with contiguous links. But every time a new link is found and inserted, the Humean question arises all over again: what is the *connection* between these intermediate causes and their effects? It seems that the Humean question never gets answered.

Oddly enough, the special theory of relativity offers a useful clue. According to that theory, light travels in a vacuum at a constant speed c, and no signal can be transmitted at any greater speed. Material particles can be accelerated to speeds approaching c, but they can never attain that speed or any greater speed.[38] It is often said, roughly, that nothing can travel faster than light, but that statement must be treated with great care. What does the "thing" in "nothing" refer to? Certainly it covers such entities as trains, baseballs, electrons, spaceships, photons, radio waves, and sound waves. But what about shadows? Are they *things* in the relevant sense of the term? The answer is negative; shadows can travel at arbitrarily high speeds. Their speed is not limited by the speed of light. Are there other examples? Yes. Consider the spot of light that is cast by a moving flashlight on the wall of a dark room. Like the shadow, this spot can travel at an arbitrarily high speed. There are many other examples. They all share the characteristic of being incapable of transmitting messages. If signals could be sent faster than light Einstein's famous *principle of relativity of simultaneity* would be undermined.

The best way to explain the difference illustrated by the examples in the preceding paragraph is, I think, to make a distinction between genuine *causal processes* and *pseudo-processes*. I shall use the unqualified term "process" in a general way to cover both the causal and pseudo varieties, but I shall not attempt to define it; examples will have to suffice. The general idea is quite simple. An event is something that happens in a fairly restricted region of spacetime. The context determines how large or small that region may be. A process is something that, in the context, has greater temporal duration than an event. From the standpoint of cosmology a supernova explosion—such as occurred in our cosmic neighborhood in the last year of the fourth decade—could be considered an event; the travel of a photon or neutrino from the explosion to earth (requiring thousands of years) would be a process. In ordinary affairs a chance meeting with a friend in a supermarket would normally be considered an event; the entire shopping trip might qualify as a process. In microphysics a collision of a photon with an electron would constitute an event; an electron orbiting an atomic nucleus would qualify as a process. A process, whether causal or pseudo, will exhibit some sort of uniformity or continuity. Something that, in one context, would be considered a single process (such as running a mile) would often be considered a complex combination of many processes from another standpoint (e.g., that of a physiologist).[39]

We must now try to distinguish between genuine causal processes and pseudo-processes. Material particles in motion (or even at rest[40]), radio waves, and photons are examples of genuine causal processes. All of them can be used to send signals; they can transmit information. Consider an ordinary piece of paper. You can write a message on it and send it through the mail to another person at another location. Or consider radio waves. They can be used to send information from a radio station to a receiver in someone's home, or from a command center to a space vehicle traveling to distant planets. Shadows and moving spots of light cast on walls are *not* causal processes; they are incapable of transmitting information. By means of another example I shall try to show why.

A lighthouse on a promontory of land sends a white beam of light that can be seen by ships at a great distance. This beam can be modified or marked. By fitting the light in the tower with a red lens, the beam becomes red. By inserting the red glass in the beam *at one place*, we can change the beam of light from white to red *from that point on*. It does not matter where in the beam the red filter is placed; wherever it is inserted it changes the beam from white to red from that point on. A genuine *causal process* is one that can transmit a mark; if the process is modified at one stage the modification persists beyond that point *without any additional intervention*.[41]

The lighthouse beacon rotates, so that it can be seen by ships in all directions. On a cloudy night, the rotating beacon casts a moving white spot on the clouds. This spot moves in a regular way; it is some sort of process, but it is *not* a causal process. If somehow the spot on the cloud is changed to red at one place—by someone in a balloon holding a piece of red cellophane right at the surface of the cloud, or by encountering a red balloon at the cloud's surface — it will suffer a modification or mark at that place, but the modification or mark will not persist beyond that place without further interventions. A 'process' that can be marked at one place, but without having any such modification persist beyond the point at which the mark is made, cannot *transmit* marks.[42] Such 'processes' are *pseudo-processes*.[43]

The moving spot of light is like the shadow in that it can travel at an arbitrarily high speed. This is easy to see. If the beacon rotates quite rapidly, its spot of light will traverse the clouds at high speed. Imagine now that the beacon continues to rotate at the same rate, but that the clouds are moved farther away. The farther away the clouds, the greater the linear velocity of the spot will be — since it has to traverse a circle of greater circumference in the same amount of time. A dramatic example is furnished by the pulsar in the Crab nebula, which rotates 30 times per second, sending out a continuous beam of electromagnetic radiation. It is about 6500 light years away. Its 'spot' of radiation sweeps by us at approximately 4×10^{13} times the speed of light.

There are many familiar examples of pseudo-processes. The scanning pattern on a cathode ray tube — a television receiver, a computer terminal, or an oscilloscope — furnishes one. The screen is made of a substance that scintillates when electrons strike it. An electron gun shoots electrons at it. As a result of the impinging electrons a spot of light moves rapidly back and forth across the screen. Another is the action as viewed at the cinema. Genuine causal processes — such as a horse running across a plain — are depicted, but what is seen on the screen is a pseudo-process created by light from a projector passing through a film and falling upon the screen. The light traveling from the projector to the screen is a whole sequence of causal processes — we might say that the light passing through each separate frame is a causal process. The motion of the horse on the screen is a pseudo-process. By momentarily shining a red light on the image of the horse on the screen it is possible to put a red spot on the horse, but if the red light does not continue to impinge on the image of the horse, the image of the horse will not continue to exhibit the red spot.[44]

The distinction between causal processes and pseudo-processes is important to our discussion of causality because processes that are *capable of transmitting marks* are processes that can also transmit information. Such processes transmit energy; they also transmit *causal influence*. They provide the *causal connections* among events that happen at different times and places in the universe.

Transmission is the key concept here. We can say, intuitively, that a process has some sort of structure — for instance, a light beam contains light of certain frequencies. A filter in its path changes that structure by removing light of some of these frequencies. Other modifications can be made by means of polarizers or mirrors. If the mark persists beyond the point in the process at which it is imposed, the structure has been modified in a lasting way *without further interventions*. If the change in structure is transmitted, the structure itself is being transmitted. Processes that can transmit marks *actually do transmit their own structure*.

Still, we should remind ourselves, the concept of transmission is a causal concept, and it must confront the Humean question. The terminology of mark transmission strongly suggests that the earlier parts of the marked process have the power to produce or reproduce the marked structure. In what does this power consist? How does the mark *get from* an earlier place in the process to a later one? Putting the question in this way suggests a strong analogy. More than 2500 years ago Zeno of Elea posed the famous paradox of the flying arrow. How does the arrow *get from* point A to point B in its trajectory? Zeno seems to have suggested that, since the arrow simply occupies the intervening positions, it must always be at rest. If it is always at rest it can never move; motion implies a contradiction.

Early in the twentieth century Bertrand Russell offered what is, I believe, a completely satisfactory resolution of the arrow paradox in terms of the so-called *at-at theory of motion*. Motion, Russell observed, is nothing more than a functional relationship between points of space and moments of time. To move *is* simply to occupy different positions in space at different moments of time. Considering any single point of space and single instant of time, an object simply is at that point of space at that instant of time. There is no distinction between being in motion and being at rest as long as we consider only that one point and that one instant.[45] To get from point A to point B consists merely of being *at* the intervening points of space *at* the corresponding moments of time. There is no further question of how the moving arrow gets from one point to another.

There was, it seemed to me, a similar answer to the question about mark transmission. A mark that is imposed at point A in a process is *transmitted* to point B in that same process if, *without additional interventions*, the mark is present at each intervening stage in the process. The difference between a process transmitting a mark and not transmitting that mark is that in the latter case the mark is present at the later stages in the process only if additional interventions occur reimposing that mark (W. Salmon 1977).

For a process to qualify as causal, it is not necessary that it actually be transmitting a mark; it is sufficient that it be capable of transmitting marks. Moreover, a causal process may not transmit every mark that is imposed upon it; a light wave that encounters another light wave will be modified at the point of intersection, but the modification will not persist beyond the point of intersection. That does not matter. The fact that it will transmit some kinds of marks is sufficient to qualify a light wave as a causal process.

With the *at-at theory of mark transmission* and the distinction between causal processes and pseudo-processes we have, it seemed to me, a satisfactory answer to Hume's basic question about the nature of causal connections. Although quite a bit more is required to develop a full-blown explication of causal explanation, this appeared to me to be the essential key (for further details see W. Salmon 1984, chap. 5).

3.8 Teleological and Functional Explanation

As we saw at the outset, a deep concern about teleology was a major impetus to the philosophical study of scientific explanation that has developed in the twentieth century. In roughly the first decade of our chronicle several of the principal early contributors to that literature – Braithwaite, Hempel, Nagel, and Scheffler – explicitly addressed the problem of teleological and/or functional explanation. This concern was instigated in large part by vitalism in biology.

Until the beginning of the third decade philosophy of biology was not pursued very actively by philosophers of science, but, as this decade opened, the situation began to change. Morton Beckner's classic, *The Biological Way of Thought*, appeared in 1968, and during the early 1970s the field began to expand dramatically with the work of such people as David Hull, Michael Ruse, and William Wimsatt. It has continued to flourish and grow right down to the present.[46] The problem of teleological/functional explanation is, of course, central to biology. Toward the end of the third decade, Larry Wright (1976) provided an account that strikes me as fundamentally correct, and it applies to many other areas in addition to biology. Eleven years later, at the very end of the fourth decade, John Bigelow and Robert Pargetter (1987) offer a theory of functional explanations that may be an improvement over Wright's, but, as we shall see, it is not *fundamentally* such a very different account.

Wright takes as his point of departure the view that teleological explanations occur frequently in science, and he maintains – contra Hempel and Beckner – that many of them are sound. He regards teleological explanations as causal in a straightforward sense – in a sense that does *not* require causes that come after their effects. At the same time, nevertheless, he considers them to be future-oriented; teleology by definition involves goal-seeking behavior. To capture both of these insights, he offers an account in terms of what he calls a *consequence-etiology*.

The basic idea of a consequence-etiology is as ingenious as it is simple. A particular bit of behavior B occurs because B has been *causally efficacious in the past* in achieving a goal G. A cat, hunting for prey, is clearly engaged in goal-directed behavior. It stalks in a typically catlike way because such stalking has resulted in the procurement of food. It is not caused by the future catching of this particular mouse, for (among other problems) in this instance he may not succeed in catching his prey. But such behavior has worked often enough to have conferred an evolutionary advantage on the members of the species. Similarly, a human being searching for water in an arid region may look for cottonwood trees, since past experience shows that they grow only when there is water in the vicinity. Such intentional behavior relies on a conviction that this action B is a suitable means of achieving G. The behavior of a homing torpedo also occurs because in the past such actions B have resulted in hitting the target (goal G). Roughly speaking, the causal efficacy of B in bringing about G in the past is, itself, an indispensable part of the cause of the occurrence of B on this occasion. It is a consequence-etiology because the consequences of doing B are a crucial part of the etiology of the doing of B.

In the case of deliberate action on the part of a human agent, it may be that direct experience of the consequence G of action B is what makes the agent do B on this occasion. Perhaps he or she has spent a great deal of time in arid regions and has found that locating cottonwood trees has been a successful strategy in finding water. Perhaps the agent has not had any such direct experience but has heard or read about this strategy. Perhaps this person has neither had direct experience, nor learned of it through reports, but, rather, inferred that it would work. Whatever the actual situation, the agent does B because he or she has reason to believe, before performing that act on this particular occasion, that B is an appropriate method for getting G.

A homing torpedo is an artifact created by a human designer to behave in a certain way. If it is being used in warfare there is a reasonable presumption (we hope) that this type of device was tested and found to perform in the desired way. The fact that similar behavior by similar objects in the past has resulted in the reaching of the goal etiologically causes the present behavior of this particular torpedo.

Examples from evolutionary biology are, I think, the clearest illustrations of Wright's model. Certain kinds of behavior, such as the stalking by a cat, become typical behavior for a species of animals because it confers an advantage with respect to the goal of survival or of reproduction. As Wright explicitly notes, the behavior may have arisen in either of two ways—deliberate creation by a supernatural agency or as a result of natural selection. The analysis of the concept of *teleology* does not tell us which of the two possible etiologies is the actual one. It only ensures that there is an etiology; empirical science must furnish the answer to which it is.

According to Wright's analysis, human conscious intentional behavior, the behavior of human artifacts that have been designed to do some particular job, and behavior that has resulted from natural selection all qualify as teleological. He offers the following schematic statement (T):

S does B for the sake of G iff:

(i) B tends to bring about G.

(ii) B occurs because (i.e., is brought about by the fact that) it tends to bring about G. (1976, 39)

The term "tends" signifies the fact that B need not *always* succeed in bringing about G. Indeed, it may be that B never brings about G, as long as it has a disposition to do so under suitable conditions. Wright offers this gloss:

> teleological behavior is behavior with a consequence-etiology: and behavior with a consequence-etiology is behavior that occurs because it brings about, is the type of thing that brings about, tends to bring about, is required to bring about, or is in some other way appropriate for bringing about some specific goal. (1976, 38–39)

Although they are obviously closely related, Wright does not identify teleological explanations with functional ascriptions or explanations. The basic reason is that only behavior is teleological—goal-directed. But in many cases something fulfills a function just by "being there." A piece of newspaper, stuffed under a door, fulfills the function of preventing a draft just because of its location. A vinyl cover lying on a playing field fulfills the function of keeping the ground dry. The function of an entity is distinguished from all sorts of other things that might result from its being where it is by the fact that its location has a consequence-etiology. The vinyl cover is where it is because it keeps the field dry. It may also catch pools of water in which children play, but that is an accidental result. Wright offers the following schematic formulation (F):

The function of X is Z iff:

    (i) Z is a consequence (result) of X's being there,
    (ii) X is there because it does (results in) Z. (1976, 81)

He comments,

> The ascription of a function simply *is* the answer to a 'Why?' question, and one with etiological force. . . . So, not only do functional explanations provide consequence-etiologies, just like explanations in terms of goals, but the simple attribution of a function ipso facto *provides* that explanation (ascription-explanation), just as does the simple attribution of a goal to behavior. This displays the enormous parallel that obtains between goals and functions, and possibly accounts for the tendency in the philosophical literature to run them together. (1976, 81)

Functional ascriptions are, of course, frequently made in physiology and in evolutionary biology. The function of the long neck of the giraffe is to enable it to reach food other animals cannot. The function of the stripes on the tiger is to provide camouflage. The function of the heart is to pump blood. The function of chlorophyll in green plants is to enable them to produce starch by photosynthesis. In the course of evolution certain attributes arise in the first instance as a result of a mutation. In such a first occurrence the attribute does not have a function, for it has no consequence-etiology at that stage. The function is established only when the recurrence or persistence of the attribute occurs because of a consequence of its previous presence.

Wright asserts explicitly that his account of teleological explanation and functional ascription is not inconsistent with a completely mechanistic explanation of goal-directed behavior (1976, 57–72). The fact that we can give a completely mechanistic account of the motion of a homing torpedo does not render it nonteleological. The fact that we can give a completely mechanical account of the operation of a governor on a steam engine does not deprive it of the function of regulating the engine's speed. He maintains, moreover—contra Charles Taylor (1964) and others—that, even if it should turn out to be possible to provide a completely physico-chemical account of the behavior of plants and animals (including the conscious behavior of human animals), that would not eliminate its teleological character. Teleological and functional explanations are on as firm ground as any explanations in any science.

The fourth-decade proposal of Bigelow and Pargetter (1987) is a "propensity theory" that they claim to be more forward-directed, temporally, than Wright's consequence-etiology. It relies heavily upon the tendency or disposition of a given item to do whatever is its function. As we have seen (schema F above), Wright refers explicitly to what the item *does*, not to what it tends to do, though on the same page, immediately after the schema, he allows for cases in which an item has a function even if it never successfully performs it. What Bigelow and Pargetter fail to note is that Wright's analysis of function supplements his analysis of teleological behavior (schema T above); like the majority of philosophers to whom Wright referred, they seem to run teleology and function together. Whether they are correct in so doing is an issue I shall not argue. The point is that Wright's analysis of teleological behavior is as much a propensity theory, and is just as forward-looking, as is the Bigelow-Pargetter theory of functions. And Wright has emphasized, as we have seen, the extremely close connection between teleology and function. I am inclined to think that, on Wright's account, every case of a function presupposes some bit of teleological behavior, but he might well disagree.[47] In any case, as Bigelow and Pargetter seem in some places to concede, their theory is not very different from Wright's. The main difference may be a disagreement over whether an item has a function the first time it occurs, or whether, as Wright maintains, it can properly be said to have a function only subsequently. If this is the *main difference*, it is a matter of fine-tuning, or as Larry used to say regarding engines of racing cars, "demon tweeking."

While I agree wholeheartedly with Wright's complete causal grounding of teleological and functional explanations, I do have one major philosophical disagreement with him. Following in the footsteps of his teacher, Michael Scriven, he declines to offer any characterization of causality itself; instead, he maintains that we have perfectly objective ways of recognizing causal relations, and that is enough. In my view, this is simply evasion of a fundamental philosophical problem. As we shall see in discussing the fourth decade, I have devoted considerable effort to the explication of causality, and I do not believe the unanalyzed notion is nearly as unproblematic as Scriven and Wright have claimed.

In 1977—the final year of the third decade—in his John Dewey Lectures, Nagel again addressed the issues of teleological and functional explanation (1977). In the course of his discussion he offers criticisms of Hempel (Nagel 1977, 305–9) and Wright (ibid., 296–301)[48] as well as others.[49] His main response to Hempel consists in a challenge to the notion of functional equivalents. According to Nagel, if one specifies with sufficient precision the nature of the organism or system within which an item has a function, it is in many cases— including Hempel's example of circulation of blood as the function of the heart in a normal human being—necessary for the fulfillment of the function. Nagel is not moved by other possible devices, such as Jarvik artificial hearts, inasmuch as they do not circulate blood in a *normal* humans. He is equally unmoved by fictitious possibilities that are not realized in nature. He therefore concludes that the normal functioning *n* in Hempel's schema entails the presence of item *i*. Nagel does acknowledge the fact that in many actual cases more than one item fulfill the same function; for example, a normal human has two ears for hearing, either one of which will do the job fairly adequately. In cases of this sort, the fulfilling of the function entails the existence of a nonempty set which may contain more than one member.

The conclusion Nagel draws from this argument is that functional explanations fit the deductive pattern. From the fact that the organism is in a normal state we can deduce the existence of the item *i* (or set of such items). If one accepts Nagel's argument about functional equivalents he overcomes the main problem that led Hempel to conclude that so-called 'functional explanations' are not real explanations at all.[50] Nagel considers that result desirable, for he maintains that many sciences, including biology in particular, do provide legitimate teleological or functional explanations. In accepting this conclusion, however, he embraces the major difficulty voiced by Braithwaite, Scheffler, and others. He allows that various kinds of facts can be explained by appeal to subsequent conditions:

> What then is accomplished by such explanations? They make explicit *one* effect of an item *i* in system *S*, as well as that the item must be present in *S* on the assumption that the item does have that effect. In short, explanations of function ascriptions make evident one role some item plays in a given system. But if this is what such explanations accomplish, would it not be intellectually more profitable, so it might be asked, to discontinue investigations of the *effects* of various items, and replace them by inquiries into the *causal* (or antecedent) conditions for the occurrence of those items? The appropriate answer, it seems to me, is that inquiries into effects or consequences are just as legitimate as inquiries into causes or antecedent conditions; that biologists as

well as other students of nature have long been concerned with ascertaining the effects produced by various systems and subsystems; and that a reasonably adequate account of the scientific enterprise must include the examination of both kinds of inquiries. (1977, 315)

No one would deny, I trust, that inquiries into effects are as legitimate and important as inquiries into causes. It does not follow from this that effects have as much explanatory import with respect to their causes as causes have with respect to their effects.

3.9 The End of a Decade/The End of an Era?

So, how did things stand at the end of the third decade? It had opened, we recall, with Hempel's (1968) attempt to tidy up the theory of statistical explanation he had offered in (1965), and it ended with his 1976 "Nachwort" (published in 1977), in which he retracted both the doctrine of *essential epistemic relativization of I-S explanation* and the *high-probability requirement*. He published no works on scientific explanation between these two. At the time Hempel admitted that there might be objectively homogeneous reference classes, I offered a serious — though flawed — attempt to explicate that concept (1977b). In relinquishing the high-probability requirement, I think, he rendered the I-S model indefensible. One could say, at the end of the third decade, that, if there is to be an admissible model of statistical explanation, its fundamental explanatory relation would be some sort of relevance relation rather than a relation of high probability of the explanans relative to the explanandum. The key doctrine of the received view — that explanations are arguments — stood on shaky ground.

At the same time there was, I think, a growing realization, for a number of reasons, that causality must play a central role in scientific explanation. To my mind, the *at-at theory* of causal influence provided a fundamental building block required for fuller elaboration of a causal theory of scientific explanation. Further development of that project occupied considerable attention in the fourth decade. As we shall see, it eventuated in the abandonment of the S-R model as an autonomous form of scientific explanation. That model became completely subservient to causal theories.

The Fourth Decade (1978–87)
A Time of Maturation

The history of the last decade is the hardest to recount, especially for one who has been deeply involved in the discussions and controversies. It is difficult to achieve perspective at such close range. Nevertheless, I think certain features are discernible. It is a period during which several different lines of thought achieved relatively high degrees of maturity. For example: (1) The role of causality in scientific explanation has been pursued in far greater detail than previously. (2) Views on the nature of statistical explanation and its relationship to causal explanation have become much more sophisticated. (3) Our understanding of the appeal to unobservables for purposes of explanation has been considerably advanced. (4) The pragmatics of explanation—which posed fundamental points of controversy from the beginning of the second decade—has been investigated more deeply and with more precision than ever before. (5) The question of the relationship between descriptive knowledge and explanatory knowledge has been examined more closely. (6) Perhaps most important, the explicandum has received significant and much-needed clarification during the decade just passed. Let us begin with this last item.

4.1 New Foundations

During the third decade of our chronicle—the one entitled "Deepening Differences"—there was an increasing awareness of the difficulties in the received view. As the hegemony crumbled philosophers looked more closely at the foundations. I recall vividly my own feeling that, in discussions of scientific explanation, the explicandum was in serious need of clarification. Too often, I felt, those who wrote on the subject would begin with a couple of examples and, assuming that these particular cases made the concept clear enough, would proceed to the task of explicating it. Referring to the exemplary job Carnap had done on clarification of the explicandum in *Logical Foundations of Probability* (chaps. I, II, and IV), I declared that the same sort of thing needed to be done for scientific explanation.

I was by no means the only philosopher to feel an urgent need for clarification of the explicandum in the mid-1970s. Michael Friedman's "Explanation and Scientific Understanding" (1974) was a seminal contribution. So also were two articles published near the close of the third decade—one by D. H. Mellor (1976), the other by Alberto Coffa (1977). Largely as a result of these latter two articles, I became acutely aware of the need to distinguish three fundamentally distinct conceptions of scientific explanation—modal, epistemic, and ontic (first presented in W. Salmon 1982).

Mellor adopts a *modal conception* of scientific explanation. Suppose some event E occurs; for all we know at the moment, it might or might not have happened. We explain it by showing that, given other circumstances, it *had to happen*. The modal conception has been advocated by a number of philosophers, but almost always in a deterministic context. If determinism is true, according to most who embrace this view, then all events are in principle explainable. If indeterminism is true, some events will not be amenable to explanation, even in principle; only those events that are necessitated by preceding conditions can be explained. In either case, however, it is not sufficient to show merely *that* the event-to-be-explained had to happen. By citing antecedent circumstances and universal laws one shows *why it had to happen*—by virtue of what it was necessitated.

But Mellor puts a novel twist on the modal conception. He claims that there can be explanations of events that are irreducibly probabilistic. By showing that an event has a high probability relative to preceding circumstances, Mellor claims, we close the gap to some extent, so to speak. There are degrees of necessitation. If an event is completely necessitated, its occurrence is fully entailed by laws and explanatory facts. If it is not fully necessitated, its occurrence is partially entailed by explanatory facts. The greater the degree of partial entailment, the better the explanation.

The concept of *partial entailment* is intuitively appealing to many philosophers who want to follow the Carnapian strategy of constructing an inductive logic on a logical interpretation of probability. The idea is to build inductive logic on the relation of partial entailment in much the same way as deductive logic can be built on the relation of full entailment. Although Carnap never used this concept in his precise explications, he does make informal reference to it (1950, 297). This concept strikes me as one of dubious value. In order to have a measure of degree of partial entailment, one is required to select an a priori measure that is tantamount to assigning prior probabilities to all of the statements that can be formulated in the language to which the inductive logic is to be applied. Given the wide range of choices for such a measure—a nondenumerable infinity (Carnap 1952)—the a priori character of this choice makes it egregiously arbitrary.[1] Thus, it seems to me, partial entailment cannot be construed as degree of necessitation in any way that is useful to the modal conception of scientific explanation. The modal conception therefore appears to require the domain of legitimate scientific explanations to be restricted to deductive explanations. Whether this restriction is tolerable in an era in which physics gives strong indication that we live in an indeterministic universe is an extremely serious question. We shall have to consider this issue in greater detail in §4.9 on deductivism. In my view it is an untenable position.

In the year following the appearance of Mellor's paper, Coffa offered a perceptive contrast between the I-S (inductive-statistical) and the S-R (statistical-relevance) models of explanation (1977).[2] On Hempel's account, we recall, an explanation—deductive or inductive—is an argument. In an I-S explanation, the relation of explanans to explanandum is an inductive or epistemic probability. In a good I-S explanation that probability is high, the higher the better. Explanations—both deductive and inductive—show that the event to be explained *was to be expected*. In view of these considerations, Coffa described Hempel's conception as *epistemic*. The S-R model is conceived in terms of objective probabilities; my formulation was given in terms of relative frequencies. In his closely related dispositional theory of inductive explanation Coffa appealed to propensities.[3] On either account, high probability per se has no particular virtue; what matters is to get the objective probabilities right. Consequently, Coffa characterized the conception underlying these models as *ontic*.

The first year of the fourth decade saw my first serious effort in print at *clarification of the explicandum*. It was embodied in "Why Ask, 'Why?'?—An Inquiry Concerning Scientific Explanation" (1978), my Presidential Address to the Pacific Division of the American Philosophical Association. Attention was focused on two basic concepts of explanation, the inferential (à la Hempel) and the causal (à la Scriven). I tried to exhibit the powerful intuitions that underlie each of them. The most appealing examples from the standpoint of the former are those in which one or more regularities are explained by derivation from more comprehensive regularities—the Newtonian explanation of Kepler's laws, or the Maxwellian explanation of the laws of optics on the basis of electromagnetic theory. Yet the vast majority of examples given by Hempel and other supporters of this conception are explanations of particular facts. Moreover—a point I did not make but should have emphasized—neither the Hempel-Oppenheim article nor Hempel's "Aspects" essay even attempts to offer an account of explanations of laws.

The most persuasive examples from the standpoint of the causal conception are explanations of particular occurrences, often in the context of practical applications. Explanations of airplane crashes and other disasters provide a wealth of instances. The main shortcoming of the causal conception, I argued, was its lack of any adequate analysis of causality. I tried to provide the foundations of such an account in terms of *causal processes, conjunctive forks*, and *interactive forks*. It was a theme to which I have returned a number of times during the last ten years.

Coffa's distinction between the epistemic and ontic conceptions of explanation arose in the context of statistical explanation. In "Why Ask, 'Why?'?" I sought to apply it more generally to the distinction between the inferential conception and the causal conception. As it seemed to me, the epistemic conception is oriented toward the notion of scientific expectability, while the ontic conception focuses upon the fitting of events into natural regularities. Those regularities are sometimes, if not always, causal. Thinking in terms of this distinction between the epistemic and ontic conceptions, I was surprised to find that, in "Aspects of Scientific Explanation," Hempel offers brief characterizations of explanation in general near the beginning and at the end. The two do not agree with each other.

In his initial discussion of D-N (deductive-nomological) explanation, Hempel says that an explanation of this type "may be regarded as an argument to the effect that the phenomenon to be explained . . . was to be expected in virtue of certain explanatory facts" (1965, 336). A bit later this conception is applied to statistical explanation as well. This characterization obviously reflects the epistemic conception. At the conclusion of this essay he sums up his theory of scientific explanation in these terms: "The central theme of this essay has been, briefly, that all scientific explanation involves, explicitly or by implication, a subsumption of its subject matter under general regularities; it seeks to provide a systematic understanding of empirical phenomena by showing that they fit into a nomic nexus" (1965, 488). For the sake of causal explanation, I would be inclined to rephrase the general characterization slightly: it seeks to provide a systematic understanding of empirical phenomena by showing *how* they fit into a *causal* nexus. Nevertheless, either way, the ontic conception is being expressed. Hempel did not, I suspect, notice the differences between his initial and final formulations.

In his doctoral dissertation — a work that is, in my opinion, quite possibly the best thing written on scientific explanation since Hempel's "Aspects" essay — Peter Railton (1980) makes an observation regarding Hempel's theory that is closely related to Coffa's distinction between epistemic and ontic conceptions. Approaching the received view in an extraordinarily sensitive way, Railton shows that Hempel's thesis to the effect that explanations explain by conferring *nomic expectability* on the explanandum cannot be maintained. The problem is that nomic expectability involves two components, *nomicity* and *expectability*, that can conflict with each other. A particular event, such as a spontaneous radioactive decay, may be rather improbable, yet we know the ineluctably statistical laws that govern its occurrence. The nomic side is fulfilled, but the expectability side is not. Hempel chose to reject, as nonexplanatory, any account that renders the event-to-be-explained improbable.[4] Railton argues that it would be better to accept such accounts as explanatory, provided they fulfill certain general conditions, and to take nomicity rather than expectability as the key to scientific explanation. If one takes that tack — as both Coffa and Railton clearly realized — it amounts to relinquishing the epistemic conception in favor of the ontic.

According to the ontic conception, the events we attempt to explain occur in a world full of regularities that are causal or lawful or both. These regularities may be deterministic or irreducibly statistical. In any case, the explanation of events consists in fitting them into the patterns that exist in the objective world. When we seek an explanation, it is because we have not discerned some feature of the regular patterns; we explain by providing information about these patterns that reveals how the explanandum-events fit in. Along with Coffa, both Railton and I endorse the ontic conception. We all maintain that explanations reveal the mechanisms, causal or other, that produce the facts we are trying to explain. The greatest difference between Railton and me concerns the degree to which explanations must be causal. His view is more lenient than mine with regard to noncausal explanation.

During the latter part of 1978 I had the great privilege of visiting Australia and offering a seminar on scientific explanation at the University of Melbourne in the Department of History and Philosophy of Science (the oldest such department in the world, I believe). This visit afforded the opportunity to think through more fully these foundational questions and to discuss them with a number of colleagues and students at Melbourne and at several other Australian universities as well. During this visit to Australia I composed the first drafts of several chapters of *Scientific Explanation and the Causal Structure of the World* (1984).

Taking the cues provided by Coffa and Mellor, it seemed to me that we could distinguish three basic conceptions of scientific explanation—modal, epistemic, and ontic—that could be discerned in Aristotle, and that have persisted down through the ages. In the context of Laplacian determinism they seem to merge harmoniously; in the context of statistical explanation they diverge dramatically. The modal conception (I claim, *pace* Mellor) precludes statistical explanation— except, perhaps, as some sort of incomplete explanation. The epistemic conception, as it occurs within the received view, requires high inductive probabilities. The ontic conception demands objective probabilities, whether high, middling, or low.[5] I offered a brief general discussion of these differing conceptions in "Comets, Pollen, and Dreams: Some Reflections on Scientific Explanation," a rather popularized account that was published in a collection of essays that grew out of my trip to Australia (Salmon 1982).

The foregoing tripartite division was a fairly serviceable crude classification scheme for theories of scientific explanation, but it required refinement. In the far more detailed treatment of the three basic conceptions (1984, chap. 4), I pointed out that the epistemic conception has three different versions—inferential, information-theoretic, and erotetic. Because of his insistence that explanations are arguments, it is appropriate to dub Hempel's conception the *inferential version* of the *epistemic conception*. This version represents the received view of scientific explanation.

Early in the third decade, James G. Greeno (1970) and Joseph Hanna (1969) began offering information-theoretic accounts of scientific explanation. Greeno took the notion of *transmitted information* as a basis for evaluating the explanatory adequacy of statistical theories. Given the fact that information is the key concept, he clearly is adopting an epistemic approach; an appropriate designation would be the *information-theoretic version* of the *epistemic conception*. A crucial difference between his model and Hempel's I-S model is that information transmitted reflects a relevance relation. Another major difference is that Greeno evaluates the explanatory power of laws or theories, but he does not provide a method for evaluating particular explanations of particular facts. This global feature constitutes an affinity between Greeno's information-theoretic account and Friedman's unification account. It seems to me that, if one wants to maintain an epistemic conception, even in the face of serious criticisms of Hempel's models, Greeno's approach is the most promising. Kenneth Sayre (1977) and Joseph Hanna (1978, 1981, 1983) have made subsequent important contributions to it.

Still another version—the *erotetic version*—of the epistemic conception must be distinguished. The term "erotetic" was chosen because the logic of questions has been known traditionally as erotetic logic. This version was suggested by Braithwaite when he remarked that "an explanation, as I understand the use of the word, is an answer to a 'Why?' question which gives some intellectual satisfaction" (1953, 348–49), but he does not develop this approach. Bromberger's work on why-questions (1966), in contrast, involves a sustained effort to elaborate the nature of why-questions and their relations to scientific explanation. But the best-known articulation of this version of the epistemic conception can be found in Bas van Fraassen's provocative work *The Scientific Image* (1980, chap. 5). This approach has obvious connections with Bromberger's earlier work, but a few major differences should be noted. First, although Bromberger presented a theory of explanations as answers to why-questions, he did not consider it a comprehensive treatment, for he flatly denied that all requests for explanations can be phrased as why-questions. Along with a number of other philosophers, Bromberger claims that some explanations are answers to (among others) how-possibly-questions, and that these are different from explanations that are answers to why-questions. Since erotetic logic deals with all kinds of questions, that fact does not disqualify Bromberger as a representative of the erotetic version of the epistemic approach. Van Fraassen, in contrast, affirms the view that all explanations can be considered answers to why-questions. Second, Bromberger made no attempt to deal with statistical explanations; van Fraassen's theory is intended to include explanations of this type. We shall discuss van Fraassen's theory in detail in §4.4.

4.2 Theoretical Explanation

Another major topic of "Why Ask, 'Why?'?" involved the appeal to unobservable entities for purposes of scientific explanation. The issue was by no means new. Within the hegemony of logical empiricism there was what might well be termed a "received view of theories." The main idea was that at the most basic level we have the particular empirical facts revealed by observation, at the next level are empirical generalizations concerning observables, and at the next higher level are theories that seem to make reference to unobservable entities. According to the received view of scientific explanation, the empirical laws explain the observed phenomena, and the theories explain the empirical laws. For example, various observed facts about pressures, temperatures, and volumes of gases are explained by the empirical ideal gas *law*, and that law is explained by the molecular-kinetic *theory*. There may, of course, be still higher level theories that explain the lower level theories. Wilfrid Sellars disparagingly dubbed this account of observable facts, empirical laws, theories, and the explanatory relationships among them the "layer cake" account. It was spelled out rather explicitly in the first four chapters of Braithwaite's *Scientific Explanation*.

During the nineteenth century there had been a good deal of resistance on the part of many scientists and philosophers to the notion that such microentities as atoms or molecules actually exist, or, at any rate, to the notion that we could possibly know anything about them if they do exist. Even those scientists who recognized the utility of the molecular-kinetic theory sometimes regarded such entities merely as useful fictions. This viewpoint is known as *instrumentalism*. The theory is a useful instrument for making scientific predictions (see Gardner 1979).

In the early part of the twentieth century these concerns about our ability to have knowledge of unobservable entities were transformed into questions about the meaningfulness of theories. Operationists and logical positivists denied that utterances putatively about unobservable objects could be scientifically meaningful. Such logical empiricists as Carnap and Hempel made serious efforts to show how—without abandoning empiricist principles—scientific theories could be meaningfully construed. An excellent account of these developments can be found in Hempel's classic essay, "The Theoretician's Dilemma" (1958). At the end of this paper he argues that theories are required for "inductive systematization," but I am not sure that he can establish more than the heuristic value of theories. Even the instrumentalist can cheerfully admit that theories are extremely useful in science. The basic question is whether the unobservable entities to which they seem to refer actually exist.

It is imperative, I believe, to separate the question of the existence of entities not directly observable by means of the unaided human senses from the issue of the meaningfulness of a theoretical vocabulary. Logical empiricists like Carnap and Hempel suggested that the terms of our scientific language can be subdivided into two parts—an observational vocabulary, containing such terms as "table," "dog," "red," "larger than," etc., and a theoretical vocabulary containing such terms as "electron," "atom," "molecule," "gene," "excited state (of an atom)," etc. The viability of a sharp observational-theoretical distinction was frequently called into question, but that particular problem need not detain us now. The instrumentalism issue can be formulated without reference to any such distinction in the scientific vocabulary.

Consider, for example, the epoch-making work of Jean Perrin on Brownian movement in the first decade of the twentieth century. To conduct his experiments he created tiny spheres of gamboge (a bright yellow resinous substance) less than a micrometer in diameter, and he accounted for their motions in terms of collisions with even smaller particles. These experiments were, as we shall see, crucial to Perrin's argument regarding the reality of molecules. Notice that I have formulated the key statements about unobservable entities without going beyond the observational vocabulary.[6]

By the time the statistical-relevance model of scientific explanation had been fairly completely articulated (circa 1970) I was aware of the fact that it was not obviously capable of accommodating theoretical explanation. At that time, James Greeno, who had developed an information-theoretic approach to statistical explanation, presented a paper (1971) in a Philosophy of Science Association symposium in which he tried to show how appeal to theories could yield an increase in information. In my comments in that symposium (1971) I showed that his approach, attractive as it was, would not work. I then set about trying to provide one that would.

In 1973 a conference on explanation (not just scientific explanation) was held at the University of Bristol; the proceedings were published in (Körner 1975). Upon receiving an invitation, but (regrettably) before writing the paper, I proposed the title "Theoretical Explanation" for my contribution. In the end I found I had written a paper on causal explanation, in which I discussed at some length appeals to continuous causal processes and common cause arguments, but which failed to yield any solid result about unobservables. In "Why Ask, 'Why?'?" I thought I had the fundamentals of an approach that would work.

Unconvinced by the various arguments about theoretical realism that had been offered by philosophers from the 1930s to the 1970s, I undertook to find out what considerations convinced natural scientists of the existence of such unobservables as atoms and molecules. Without having been aware at that time of the historical importance of Avogadro's number N, I did recognize that it provided a crucial link between the macrocosm and the microcosm. With the help of N, one could calculate micro-quantities from macro-quantities and conversely. From the mass of a mole of any given substance, for example, N gives us immediately the mass of a molecule of that substance. Thus, it seemed to me, the ascertainment of N was a good place to start.

A first and most obvious way to get at the value of N is through the phenomenon of Brownian movement. According to the molecular-kinetic theory of gases, as Einstein and Smoluchowski showed in 1905–06, the motion of a Brownian particle suspended in a gas is the result of random bombardment by the molecules of the gas. Assuming that the gas and the Brownian particles are in thermal equilibrium, it follows that the average kinetic energy of the Brownian particles is equal to the average kinetic energy of the molecules. As Perrin remarked at about the same time, the dance of the Brownian particles represents qualitatively the random motion of the molecules. By ascertaining the mass of the Brownian particle, the average velocity of the Brownian particle, and the average velocity of the molecules, one can compute the mass of the molecule directly. In practice the situation is a bit more complicated. Although the mass of the Brownian particle and the average velocities of the molecules are quite directly measurable, the average velocity of the Brownian particle is not, for it changes its direction of motion too rapidly. But by indirect means—basically, observation of rates of diffusion of Brownian particles—what amounts to the same method can be applied. As a result we know the mass of a molecule of a gas and Avogadro's number N, the number of molecules in a mole of that gas. This type of experiment was done in the early years of the twentieth century by Perrin.[7]

The instrumentalist can easily reply to the preceding consideration by pointing out that since the advent of the molecular-kinetic theory we have known that it is useful to think of a gas as composed of little particles that move at high speeds and collide with one another and with the walls of the container. Now, it can be added, the Brownian particle behaves as if it is being bombarded by these little particles, and, indeed, we can say that the gas behaves as if it is composed of a certain number of these tiny particles. But all of this does not prove the reality of molecules; it shows that the molecular kinetic theory is an excellent instrument for predicting the behavior of gases (and of the Brownian particles suspended in them). Thus, the manifest success of the molecular-kinetic theory did not constitute compelling evidence for the existence of molecules. As a matter of historical fact many serious and knowledgeable physical scientists at the turn of the century did not believe that atoms and molecules are real.

The way out of this difficulty lies in the fact that N can be ascertained in a variety of ways. In "Why Ask, 'Why?'?" I mentioned a determination by means of electrolysis as an example of a totally different experimental approach. If an electric current passes through a solution containing a silver salt, an amount of metallic silver proportional to the amount of electric charge passing through the solution is deposited on the cathode. The amount required to deposit one mole of a monovalent metal (such as silver) is known as a faraday. A faraday is found by experiment to be 96,487 coulombs. If that charge is divided by the charge on the electron, empirically determined by J. J. Thomson and Robert Millikan to be -1.602×10^{-19} coulombs, the result is N. A faraday is simply Avogadro's number of electron charges. Superficially, the phenomena involved in these two experiments are entirely different, yet they agree in the numerical value they yield for N. I suggested that this agreement, within experimental error, of the value derived via the study of Brownian movement with the value derived via the electrolysis experiment suggests that the particulate character of matter—the reality of such things as molecules, atoms, ions, and electrons—is a common cause of this agreement. I realized at the time that there are many other ways of ascertaining N, but mentioned just two in order to illustrate the common cause character of the argument.

At some time during 1978 I became aware of Mary Jo Nye's superb historical account of the work of Jean Perrin (1972) and, through that, of Perrin's own semi-popular treatment (1913, English translation 1923). In the period between 1905 and 1913 Perrin had done a spectacular set of experiments on Brownian movement and the determination of N. Combining his findings with those of other workers, Perrin presents a table, near the close of his book, listing the results of *thirteen distinct methods* for ascertaining N, and notes the striking agreement among them. Immediately after the table he remarks:

> Our wonder is aroused at the very remarkable agreement found between values derived from the consideration of such widely different phenomena. Seeing that not only is the same magnitude obtained by each method when the conditions under which it is applied are varied as much as possible, but that the numbers thus established also agree among themselves, without discrepancy, for all the methods employed, the real existence of the molecule is given a probability bordering on certainty. (1923, 215–16)

This was, indeed, the argument that convinced virtually every physical scientist of the existence of unobservable entities of this sort.

In *Scientific Explanation and the Causal Structure of the World* (213–27) I tried to spell out this common cause argument in some detail. However, instead of dealing with thirteen different ways of ascertaining N, I confined my attention to five: Brownian movement, electrolysis, alpha radiation and helium production, X ray diffraction by crystals, and blackbody radiation. They constitute a highly diverse group of experiments. I thought it fitting, given St. Thomas Aquinas's five ways of demonstrating the existence of God, that I should cite the same number to establish the existence of atoms and molecules. I consider the argument that convinced the scientists in the early part of the twentieth century philosophically compelling.

4.3 Descriptive vs. Explanatory Knowledge

What good are explanations? This question has been asked—explicitly or implicitly—on innumerable occasions over the years, and it has received a variety of answers. During the first half of the twentieth century, scientific philosophers were concerned to refute such answers as, "Explanations inform us of the ultimate purposes for which the world was created," "Explanations reveal the underlying essences of the things we find in the world," or "Explanations exhibit the vital forces or entelechies within living beings." As I remarked near the beginning, the heavy theological and metaphysical involvements of philosophy during that period led some scientists and philosophers to reject altogether the notion that science has anything to do with explanation. Those who did not want to relinquish the claim that science can furnish legitimate explanations were at pains to make it clear that explanatory knowledge is part of our empirically based descriptive knowledge of the natural world. The classic paper by Hempel and Oppenheim, as well as Hempel's "Aspects" paper, were efforts to delineate exactly what sort of descriptive knowledge constitutes explanatory knowledge. These essays were designed to show that legitimate scientific explanations could be had without appealing to superempirical facts or agencies. This is what the received view was all about.

At the same time, these treatments of explanation rejected such psychologistic answers as "Explanations increase our understanding by reducing the unfamiliar to the familiar," or "Explanations ease our intellectual discomforts and make us feel more at home in the world" (see H-O 1948, §4; W. Salmon 1984, 12–15). Great caution must be exercised when we say that scientific explanations have value in that they enable us to understand our world, for *understanding* is an extremely vague concept. Moreover—because of the strong connotations of human empathy the word "understanding" carries—this line can easily lead to anthropomorphism. The received view was concerned also to show that scientific explanations can be given without indulging in that intellectual vice either.

These considerations lead, however, to a deeply perplexing puzzle. If explanatory knowledge does not exceed the bounds of descriptive knowledge, in what *does* it consist? What do we have when we have explanations that we did not already have by virtue of our descriptive knowledge? What is the nature of the understanding scientific explanations are supposed to convey? The full force of this problem did not strike me until about 1978, and I addressed it explicitly in "Why Ask, 'Why?'?" (1978). At about the same time, van Fraassen was asking the same question. We shall consider his answer, which is quite different from mine, below.

The way I posed the problem is this. Suppose you were Laplace's demon, possessing a complete description of the world at one particular moment, knowing all of the laws, and having the ability to solve the mathematical problems required to predict or postdict everything that ever has happened or ever will happen. Such knowledge would appear to be descriptively complete; what else would be involved in having explanations?

I now think this was a particularly inept way to put the question, for one obvious answer is that the demon would have no occasion to ask "Why?" Van Fraassen would, I suspect, agree with this response. As our discussion of Bromberger's work has revealed, one asks a why-question only as a result of some sort of perplexity—recall his p- and b-predicaments. The demon would not be in any such predicament. So we should drop the fantasy of the Laplacian demon. Why-questions are raised only in contexts in which our knowledge is incomplete. Nevertheless, the fundamental question remains, and remains important. What sort of information is explanatory information? How, if at all, does explanatory knowledge differ from other types of descriptive knowledge? The three basic conceptions discussed in §4.1 offer distinct answers to this question.

For the modal conception a straightforward answer is available. Explanatory knowledge adds a modal dimension to our descriptive and predictive knowledge. Explanatory knowledge is knowledge of what is necessary and what is impossible.[8]

In "Why Ask, 'Why?'?" I attempted to respond on behalf of the ontic conception. Even though the question was badly put, the answer points, I think, in the right direction. I suggested that, in addition to purely descriptive knowledge, one would need causal knowledge: recognition of the difference between causal and noncausal laws, the difference between causal processes and pseudo-processes, and the difference between causal interactions and mere spatio-temporal coincidences. At present I would be inclined to phrase the answer somewhat differently in terms of laying bare the underlying mechanisms, but the basic idea seems to me sound. According to the ontic conception, explanatory knowledge is knowledge of the causal mechanisms, and mechanisms of other types perhaps, that produce the phenomena with which we are concerned.

Since it comes in three distinct versions, the epistemic conception requires three answers. According to the received view, explanations involve the subsumption of the fact to be explained under some kind of lawful regularity, universal or statistical. In that way they provide *nomic expectability*. But, one might ask, do we not already have nomic expectability as part of our descriptive and predictive knowledge?[9] The received view seeks assiduously to avoid any incursion into metaphysics or theology for purposes of scientific explanation, but it is hard to see just what constitutes the explanatory import of an explanation on that conception. Indeed, given the original strong version of Hempel's thesis of the symmetry between explanation and prediction, this point obviously holds. The *only* difference between explanation and prediction is pragmatic. If we know that the fact described in the conclusion of an argument conforming to one of the models of explanation actually obtains, the argument constitutes an explanation of that fact. If we do not already know that this fact obtains, the very same argument constitutes a prediction.

As Israel Scheffler pointed out around the beginning of the second decade, it is highly implausible to maintain that every legitimate scientific prediction can serve, under suitable pragmatic conditions, as an explanation. One major reason, as we saw, is that inductive arguments from particulars to particulars—not involving any law—can provide reliable predictions. The symmetry thesis must therefore be amended, at the very least, to the claim that predictions based upon laws can function as explanations in certain epistemic contexts. Thus, the appeal to *nomic regularities* is the crucial feature of explanation. Since it seems reasonable *prima facie* to claim that law statements constitute a crucial part of our descriptive knowledge, we still must ask in what way explanations, as characterized by the received view, have *explanatory import*.

It is useful to recall, in this connection, the fundamental point made by Railton concerning nomic expectability—namely, sometimes nomicity and expectability conflict. If one goes with expectability, thus fully preserving the epistemic character of the conception, descriptive and explanatory knowledge seem indistinguishable. If one opts for nomicity instead, one ends up—as Hempel did at the conclusion of "Aspects"—in the ontic conception.

Before going on to consider the other two versions of the epistemic conception, let us pause to compare the three responses already given. When we appeal to modality—to necessity or impossibility—it may involve either of two distinct approaches. First, one might look upon the modalities as metaphysical categories, residing in a realm separate from the domain that can be empirically investigated. This construal would exile explanation from science. Second, one might say that nothing beyond physical necessity and impossibility is involved, and that these come directly from the laws of nature. If law-statements constitute part of our descriptive knowledge, then the modal conception makes no appeal to anything beyond our descriptive knowledge. Notice that, on this interpretation, the modal conception and the received view are in complete agreement on the status of D-N explanation. The received view differs from the modal conception only insofar as it admits some form of statistical explanation.

When we look at the inferential version of the epistemic conception—involving *nomic* expectability—we see immediately that the status of laws becomes crucial. Consequently, we must ask, what is the difference between true lawlike generalizations (of either the universal or statistical variety) and true accidental generalizations? Recall our previous example comparing statements about massive spheres of gold with statements about massive spheres of uranium. Is there any objective difference between lawful and accidental generalizations, or is it merely a matter of our greater confidence in one as opposed to the other? Alternatively, do we want to characterize laws in terms of relations among universals? If so, this takes the distinction between laws and nonlaws, and, consequently, the characterization of explanation, out of the realm of empirical science. It appears, therefore, that the choice, for the adherent of the received view, is between a heavily pragmatic theory (such as we found in Rescher) and an extrascientific metaphysical one. Neither choice, I believe, captures the intent of the received view.

According to the ontic conception, there is a further gap between explanation and prediction. As we noted much earlier in connection with the famous barometer example, the sharply falling barometric reading is a satisfactory basis for predicting a storm, but contributes in no way to the explanation of the storm. The reason is, of course, the lack of a direct causal connection. For the ontic conception, therefore, mere subsumption under a law is not sufficient for explanation. There must be, in addition, a suitable causal relation between the explanans and the explanandum—at least as long as we steer clear of quantum mechanical phenomena.

The basic question, for this conception, is the status of causality. If we construe causal relations in extrascientific metaphysical terms we will banish explanation from science. If we follow a purely Humean tack, construing causality strictly in terms of constant conjunction (see Mackie 1974),[10] we will make the ontic approach identical with the received view.[11] The approach I adopted in "Why Ask, 'Why?'?" involved an appeal to causal processes and causal interactions (see §3.6 above). Causal processes are distinguished from pseudo-processes in terms of the ability to transmit marks. I attempted to give an entirely empirical construal of the causal concept of transmission in terms of the *at-at* theory. I distinguished causal interactions from mere spatio-temporal intersections of processes in terms of mutual modification of the processes involved. To my great regret, I found no way of carrying out these explications without the use of counterfactual conditionals (W. Salmon 1984, 148–50, 171–74). I have no analysis of counterfactuals, though I do offer a method for testing the truth of the kinds of counterfactuals that are invoked in this context. I am not terribly dissatisfied with this state of affairs, but doubtless other philosophers will not be as easily satisfied on this score.

We have known for a long time that three sets of issues are tightly intertwined: modalities, laws, and counterfactuals (see W. Salmon, 1976). It seemed that any two of the three could be satisfactorily analyzed in terms of the third, but it is difficult to produce a satisfying analysis of any one of them that did not invoke at least one of the other two. Circularity is, consequently, always a serious threat. The most popular current approach to counterfactuals seems to be one that appeals to possible worlds (see Lewis 1973).[12] It has two major shortcomings in my opinion. First, the postulation of the existence of myriad possible worlds, distinct from our actual world, takes us deep into the superempirical.[13] Second, evaluation of the similarity of possible worlds—which is essential to the analysis of counterfactuals—requires an appeal to laws. So, it has not broken us out of the circle. It is quite interesting to note, then, that (1) the modal conception attaches itself to modality, (2) the epistemic conception, inferential version, depends upon laws, and (3) the ontic conception—as far as I can see, at least—appeals to counterfactuals. Each conception has its cross to bear. Yet in that cross may lie the 'something extra' beyond sheer descriptive knowledge that is supposed to provide scientific understanding.

Let us now return to the remaining two versions of the epistemic conception. The information-theoretic version faces almost the same difficulty as did the inferential version in explicating the distinction between purely descriptive knowledge and explanatory knowledge. According to the information-theoretic approach the explanatory value of a law or theory is measured in terms of its efficacy in improving predictive power. That is just what information transmitted amounts to. There is, nevertheless, a possible approach that can be shared by the inferential version and the information-theoretic version. In discussing explanation, prediction, the status of theories, and other related concepts, Hempel has often referred generally to *systematization* — both *deductive systematization* and *inductive systematization*. This suggests the possibility of construing explanatory force, not in terms of some extra type of knowledge, but rather, in terms of the organization of the descriptive knowledge we have or can procure.

This way of looking at explanatory force bears a striking resemblance to Friedman's unification theory of explanation. According to that view, explanations improve our understanding through the unification of our knowledge. Our understanding is increased when we can reduce the number of independent assumptions we have to make about the world. Another way to look at unification is in terms of information theory. When a great deal of information about the world is contained in a short message, we have increased understanding. Either way, what is crucial to explanation is not some particular kind of explanatory knowledge, but, rather, the way in which our descriptive knowledge is organized.

According to the erotetic version of the epistemic conception, as developed by van Fraassen, there is no fundamental difference between descriptive knowledge and explanatory knowledge. On van Fraassen's theory an explanation is simply an *answer* to a why-question; it is nothing other than descriptive information that, *in a given context*, answers a particular type of question. Whether a piece of information constitutes explanatory knowledge depends solely upon the context in which it is furnished. Thus, whatever distinction there is between descriptive and explanatory knowledge is entirely pragmatic.

When we use the term "descriptive knowledge," there is serious danger of equivocation. In one meaning of the term, describing the world or any part of it consists only in reporting what is apparent to the senses. Thus, one could describe a given volume of air as warm or cold, calm or windy, clear or hazy, etc. One might also include such readily detectable features as moisture content and pressure. Such a description would *not* include the fact that air consists of molecules of various gases such as nitrogen, oxygen, and carbon dioxide, or that these molecules are made up of atoms having certain characteristics. A complete description of this sort would be a description solely of appearances. In another sense, a complete description of an entity would include all kinds of facts about it, directly observable or not. Such a description of the same body of air, if complete, would specify how many molecules there are of each type, the atomic constitution of molecules of each type, and facts about the collisions of molecules. This is the sort of description Laplace's demon would have possessed. When we raise the question about the relationship between descriptive knowledge and explanatory knowledge, we must be careful to indicate in which of the foregoing two senses we are construing the word "descriptive." This has a crucial bearing upon the theory of scientific explanation.

In the first chapter of his doctoral dissertation, Coffa discussed in detail the nature of explication, and he devoted serious attention to the clarification of the explicandum he was trying to explicate. He called attention to

> a centuries-old philosophical tradition, sometimes referred to by the name of 'instrumentalism', that has denied the claim that science has explanatory power. For instrumentalists there are no scientific explanations. Science is acknowledged to have a number of virtues, but none of them is associated with the production of a better understanding of what goes on in the world. Science could be a source of predictive power, the foundation of technology, an unlimited fountainhead of aesthetic pleasure, the maximal organizational principle of experience; but it would not be a source of understanding. For the instrumentalist science is a prediction machine with pleasant, but only psychological side-effects.
>
> The notion of explanation we want to explicate is one whose correct applicability to a scientific argument is denied by instrumentalists. An alleged theory of explanation (i.e., an explication of explanation) that elucidates the concept of explanation in such a way that the instrumentalist may consistently agree that science contains explanations in the explicated sense, will not be an explication of any of the explicanda described here. (1973, 61–62)

Coffa's main object, in raising the instrumentalism-realism issue was to apply it to the received view. According to Hempel's account, a deductive-nomological explanation is an argument that contains essentially a lawful generalization. This law may be a generalization concerning observables only—for example, our old friend the ideal gas law. The subsumption of a fact under that generalization constitutes a satisfactory explanation on Hempel's account. If we go on to explain the ideal gas law in terms of the kinetic-molecular theory, that is a distinct explanation, and its existence does nothing to impair the status of the foregoing explanation. As the received view has it, there are legitimate scientific explanations that do not appeal to theories about unobservables; indeed, it is possible in principle (in a world very different from ours), though contrary to fact, for all legitimate explanations to be given wholly in terms of laws governing observable phenomena.[14] Thus, it is possible in principle for the instrumentalist to embrace the received view without qualms. To Coffa's mind, this constitutes a severe shortcoming of any account of scientific explanation. I agree.

To see how these considerations apply to the ontic conception, let us focus upon causal explanation. The Humean tradition suggests that the world is full of constant conjunctions among observable phenomena, but that there is serious doubt as to whether there is anything in the external world that distinguishes accidental conjunctions from genuine causal relations. Hume seems to say that the distinction is in the mind — in the imagination. I do not believe that causality resides in the mind; moreover, I think it is a distinctly nontrivial matter to distinguish genuine cause-effect relations from accidental correlations or from cases in which the correlation is produced by a common cause. Furthermore, in many cases in which we seek a cause we do not even have a constant conjunction *among observables* to appeal to. For example, the nearly simultaneous illness of a number of people who had attended the American Legion convention in Philadelphia in 1976 could not be satisfactorily explained until the *Legionella* bacillus was identified, and its source in Philadelphia located.

As I said above, in "Why Ask, 'Why?'?" I attempted to spell out in some detail the nature of the causal mechanisms that seem to exist in our world. They include causal processes, causal interactions (interactive forks), and conjunctive forks. Our casual observation of phenomena seldom reveals the causal mechanisms; indeed, careful scrutiny of the observable phenomena does not generally reveal their presence and nature. So our efforts at finding causal relations and causal explanations often — if not always — take us beyond the realm of observable phenomena. Such knowledge is empirical knowledge,[15] and it involves descriptive knowledge of the hidden mechanisms of the world, but it does go beyond descriptive knowledge of the observable phenomena. There is no *logical* necessity in the fact that causal mechanisms involve unobservables; that is just the way our world happens to work.

As we have already noted, Coffa is a staunch defender of the ontic conception of scientific explanation, and his theory of explanation reflects this attitude. An explanation of any occurrence is a set of objective facts and relationships. For Coffa, what explains an event is whatever produced it or brought it about. Explanations are nomic dispositions of universal (p = 1) or less than universal (p < 1) strength. The linguistic entities that are often called 'explanations' are statements reporting on the actual explanation. Explanations, in his view, are fully objective and, where explanations of nonhuman facts are concerned, they exist whether or not anyone ever discovers or describes them. Explanations are not epistemically relativized, nor (outside of the realm of human psychology) do they have psychological components, nor do they have pragmatic dimensions. Since the mechanisms that operate in our world are frequently hidden, the true explanation (on Coffa's view) is often something whose existence the instrumentalist denies.

Traditionally, instrumentalism has been opposed to one sort of realism or another. The instrumentalist in physics recognizes that the molecular-kinetic theory of gases provides a useful tool for establishing regular relationships among such observables as temperature, pressure, volume, and mass, but denies that such things as molecules actually exist. The instrumentalist in psychology sees the function of psychological theories as providing relationships between observable stimuli and observable responses, without appealing to such unobservable entities as feelings of hunger, anxiety, or pain.[16] For the instrumentalist, our descriptive knowledge of the world is confined to knowledge of whatever is more or less directly observable. The instrumentalist cannot appeal to unobservables for purposes of explaining observed fact, for he or she denies that any such things exist.

The realist, in contrast, makes at least a rough demarcation between descriptive knowledge of observables and descriptive knowledge of unobservables. According to the realist, we can, in principle, have knowledge of observables by direct observation—though, in fact, much of our knowledge even of observables is rather indirect. The color, shape, size, and surface texture of a satellite of Pluto, for instance, can at present be known only by means of complex theoretical inferences. In addition, the realist claims, we can have descriptive knowledge of unobservables on the basis of theoretical inferences. Although some realists might maintain that our alleged knowledge of unobservables transcends the empirical realm, I claim that we have fully empirical knowledge of them. My thesis is that realism and empiricism are entirely compatible, for we can confirm or disconfirm theories about unobservables on the basis of observational evidence.[17] The realist constructs explanatory theories that are intended and believed to make reference to unobservable entities. The realist asserts that such things as atoms, molecules, ions, subatomic particles, and microorganisms actually exist, and that we can explain a vast range of physical, chemical, and biological phenomena in terms of their behavior. As we shall see, this claim is sharply denied by van Fraassen, who is *not* an instrumentalist, but who shares a number of important views with philosophers of that persuasion.

For the proponent of the ontic conception of scientific explanation, realism provides a straightforward answer to the question of the distinction between descriptive and explanatory knowledge. Taking "description" in the narrower sense which includes only description of appearances, the realist can say that explanatory knowledge is knowledge of the underlying mechanisms—causal or otherwise—that produce the phenomena we want to explain. To explain is to expose the internal workings, to lay bare the hidden mechanisms, to open the black boxes nature presents to us.

The foregoing discussion of the relationship between descriptive and explanatory knowledge has focused almost exclusively on the intellectual value of scientific explanations. We should not forget that explanations have practical value as well. Finding scientific explanations of various types of occurrences often points to useful ways of controlling important features of our world. It may help to eliminate such undesirable events as epidemics and airplane crashes. It may help to bring about such desirable results as a greater healthy life span for humans. It may also help to alleviate superstitious fears.[18] But, to return to our main theme, our discussion has left us with three apparently viable answers to the question of the intellectual value of scientific explanations. Such explanations enhance our understanding of the world. Our understanding is increased (1) when we obtain knowledge of the hidden mechanisms, causal or other, that produce the phenomena we seek to explain, (2) when our knowledge of the world is so organized that we can comprehend what we know under a smaller number of assumptions than previously, and (3) when we supply missing bits of descriptive knowledge that answer why-questions and remove us from particular sorts of intellectual predicaments. Which of these is *the* function of scientific explanation? None *uniquely* qualifies, I should say; all three are admissible. But not everyone would agree. And even among those who do agree, some will say that one of the three is fundamental, and the others have a distinctly derivative status.

4.4 The Pragmatics of Explanation

The most articulate and prominent anti-realist of the fourth decade is Bas van Fraassen. The full statement of his position appears in *The Scientific Image* (1980), but it had been anticipated to some extent in articles that appeared near the close of the third decade (1976, 1977). It should be carefully noted that van Fraassen is *not* an instrumentalist; his position is *constructive empiricism*. Like a sophisticated instrumentalist, van Fraassen recognizes that theories, which at least appear to make reference to unobservable entities, have played an indispensable role in the development of modern science, and he recognizes that in all likelihood they will continue to do so. The most fundamental difference between the instrumentalist and the constructive empiricist is that the former *denies* the existence of unobservables while the latter remains *agnostic* with respect to their existence.

According to van Fraassen, accepted scientific theories are *accepted as empirically adequate*, but they need not be *believed to be true*. Thus, he claims, when we accept a theory that seems to make reference to unobservables for various purposes—including use in giving explanations—we are committed to claiming that it yields true statements about observables, but we are not committed to claiming that what it says about unobservables is true. Likewise, however, we are not committed to claiming that what it says about unobservables is false. Van Fraassen's thesis is that one cannot be convicted of irrationality for disbelief in such things as molecules and one cannot be convicted of irrationality for belief in them. Regarding the name he has chosen for his position, he says, "I use the adjective 'constructive' to indicate my view that scientific activity is one of construction rather than discovery: construction of models that must be adequate to the phenomena, and not discovery of truth concerning the unobservable" (1980, 5).

In spite of his anti-realism, van Fraassen offers a theory of explanation that fits with his overall conception of the nature of science. Contrasting his view with the more traditional approaches (including, of course, the received view), he says,

The discussion of explanation went wrong at the very beginning when explanation was conceived of as a relationship like description: a relation between theory and fact. Really it is a three-term relation, between theory, fact, and context. No wonder that no single relation between theory and fact ever managed to fit more than a few examples! (1980, 156)

This statement contains a complete rejection of the conception upon which Coffa focused his attention, an ontic conception that located explanations in the external world and which is totally unavailable to instrumentalists.

The theory of explanation van Fraassen offers is *not* intended to be a theory only of scientific explanation; it should encompass other kinds of explanations as well. On his view, an explanation is *an answer to a why-question*. A scientific explanation is one that relies essentially on scientific knowledge. We might ask, for example, why Hitler invaded Russia during World War II. It is a historical question, and it calls for a historical answer. Whether or not history is a science, an answer to this question will be an explanation, and it will be within the purview of van Fraassen's theory.[19] Similarly, one might ask why the cat is sitting in front of the door. The common-sense answer is that he wants to go outside. Again, van Fraassen's theory is meant to handle such questions and answers. History and common sense are, after all, closely related to science whether or not they qualify as parts of science. It is not too unreasonable to expect explanations in those domains to resemble scientific explanations.

There are, however, why-questions that do not seem to be calls for explanations. One might ask, in a time of grief, why a loved one had died. Such a question is not intended to evoke an explanation; it is a cry for sympathy. Exclamations and tears may constitute a far better response than would any factual proposition. Indeed, the grieving individual may be fully aware of the scientific explanation of the demise. Other why-questions seem best interpreted as requests for moral justification. The question has been raised in courts of law as to why a member of a minority group was admitted to medical school to the exclusion of some non-minority candidate whose qualifications were somewhat better. The point at issue is the ethical basis for that decision. It appears that many why-questions are not requests for explanations; consequently, to sustain the claim that explanations are answers to why-questions we would have to distinguish *explanation-seeking* why-questions from other kinds of why-questions. Van Fraassen does not undertake that task, but one might reasonably claim that contextual cues, which play a central role in his theory, should enable us to sort them out.

Given that not all why-questions seek to elicit explanations, the next question is whether all explanations—at least, all scientific explanations—are sought by means of why-questions. Clearly the answer to this question is negative. It has been suggested, however, that any request for a scientific explanation—no matter how it is actually formulated—can be appropriately rephrased in the form of a why-question. Inasmuch as van Fraassen claims that explanations are answers to why-questions, he is patently committed to this view. A number of philosophers have denied it; for instance, Bromberger, William Dray, and Frederick Suppe have argued that some explanations are answers to questions of *how-possibly*. Dray had raised this issue (1957) and it is discussed by Hempel (1965, 428–30).

Consider a concrete example. There is an old saying that when a cat falls (from a sufficient height) it always lands on its feet. We know, however, that angular momentum is conserved. How is it then possible for a cat, dropped from an adequate height with zero angular momentum and with its legs pointing upward, to land on its feet? Your first reaction might be to suppose that the old saying is simply not true—that it is an old wives' tale. But that is not correct. Experiment has shown that the cat can twist its body in various ways while the net angular momentum remains zero to achieve the desired position upon landing.[20] A diver who does a twist, as distinguished from a somersault, achieves a similar feat. Hempel's response to such examples is to admit that there is a pragmatic difference between why- and how-possibly-questions, in that the person who poses a how-possibly question is under the mistaken impression that the occurrence is either physically impossible or highly improbable. The appropriate response is to expose the misapprehension and produce either a D-N or an I-S explanation of the phenomenon in question. The original question could thus have been rephrased, "Why did this event (which I initially regarded as impossible or highly improbable) occur?" (Hempel 1965, 428–30).

In (1984, 10) I also expressed the claim that all requests for scientific explanations can be formulated as why-questions, but I now suspect that it is mistaken. Hempel's response seems inadequate for two reasons. First, a how-possibly question does not require an actual explanation; any potential explanation not ruled out by known facts is a suitable answer. For example, a DC-9 jet airplane recently crashed upon takeoff at Denver's Stapleton Airport during a snowstorm. One peculiar feature of this accident is that the plane flipped over onto its back. There are many explanations of a crash under the circumstances, but I wondered how it could have flipped over. Two how-possibly explanations were mentioned in the news reports. One is that it encountered wing-tip turbulence from another airplane just after it became airborne. Another was suggested by the report of a survivor, who claimed that the plane was de-iced three times during its wait for departure, but that on the latter two of these occasions one wing, but not the other, was treated. If one wing had an accumulation of ice on its leading edge while the other did not, the difference in lift provided by the two wings might have been sufficient cause for the plane to flip over. As I write this paragraph I have not yet heard the final determination regarding the cause of this crash. Both potential explanations I have mentioned are satisfactory answers to the how-possibly question, but we do not know the correct answer to the why-question.

Second, improbable events do occur. Not long before the Denver crash, another DC-9 crashed on take-off at Detroit's Metropolitan Airport. Investigators have concluded, I believe, that the pilot failed to extend the wing flaps for take-off. It is extremely unlikely that an experienced pilot would make such an error, that the co-pilot would fail to notice, and that the warning signal would fail to be sounded or would be ignored. But apparently that is what happened. Hempel suggests that we are obliged to find additional factors that would make the errors highly probable. It seems to me that, even if one insists on high probabilities for explanations that answer why-questions, no such thing is required for answers to how-possibly questions. It is sufficient to show that the probability is different from zero.

Still other explanations may be answers to *how-actually* questions. How did there come to be mammals (other than bats) in New Zealand? They were humans, who came in boats, and who later imported other mammals. This is, I think, a genuine scientific explanation. It is not an explanation of why they came; rather, it is an explanation of how they got there.

Having posted some caveats about simply identifying explanations as answers to why-questions, I shall now turn to a discussion of some of the details of van Fraassen's theory of explanation. Before doing so, it is worth noting that, although this issue is crucial for an advocate of the erotetic conception, it has little — if any — genuine significance for the proponent of the ontic conception. According to this latter conception the search for explanations is a search for underlying mechanisms, and the form of the question requesting them is not very important.

The development of van Fraassen's theory of why-questions and their answers in the fourth decade was greatly facilitated by progress in formal pragmatics and the publication, near the end of the third decade, of Belnap and Steel's *The Logic of Questions and Answers* (1976). Although this is a landmark work in erotetic logic, it contains hardly any treatment of why-questions. To begin, it is essential to realize that, as van Fraassen sets forth his theory, questions and answers are abstract entities. An answer is a proposition, and a given proposition can be expressed by means of many different declarative sentences. Moreover, a particular sentence, uttered on different occasions, can express different propositions. For any given sentence, the context determines which proposition it expresses. "I am here," for example, always expresses a true proposition, but the proposition it expresses depends on who utters the sentence, and where and when. Similarly, a given question may be expressed by many different interrogative sentences, and a particular interrogative may pose different questions on different occasions. When an interrogative sentence is uttered the context determines which question is being asked.

According to van Fraassen, we can think of the standard form of a why-question as

Why (is it the case that) P_k? (Q)

where P_k states the fact to be explained (the explanandum phenomenon). Such a question can be identified with an ordered triple

$$Q = <P_k, X, R>$$

where P_k is the *topic* of the question, $X = \{P_1, P_2, \ldots, P_k, \ldots\}$ is the *contrast class*, and R is the *relevance relation*. To take a familiar example, consider the question,

Why did the Bunsen flame turn yellow?

The topic is

The Bunsen flame turned yellow.

The contrast class is

The Bunsen flame remained blue (P_1)
The Bunsen flame turned green (P_2)
The Bunsen flame turned orange (P_3)

.

.

.

The Bunsen flame turned yellow (P_k)

.

.

.

The relevance relation R is the relation of cause to effect. Hempel's answer, we recall, is that a piece of rock salt was placed in the flame, rock salt is a sodium compound, and all sodium compounds turn Bunsen flames yellow.

A crucial feature of van Fraassen's account is its emphasis upon the fact that the same interrogative sentence – the same group of words – can express different questions. This can easily happen if different contrast classes are involved. He invites consideration of the interrogative

Why did Adam eat the apple?

By the inflection or emphasis of the speaker, or by other contextual clues, we might find that any of three different questions is being expressed. It might mean,

Why did *Adam* eat the apple?

where the contrast class = {Eve ate the apple, the serpent ate the apple, the goat ate the apple, etc.}. At the same time, it might mean

Why did Adam *eat* the apple?

where the contrast class = {Adam ate the apple, Adam threw the apple away, Adam gave the apple back to Eve, Adam fed the apple to the goat, etc.}. Also, it might mean

Why did Adam eat the *apple*?

where the contrast class = {Adam ate the apple, Adam ate the pear, Adam ate the pomegranate, etc.}. The context determines which is the appropriate contrast class.

Another feature of questions is that they generally come with *presuppositions*, and why-questions are no exception. The presupposition of Q is

(a) P_k is true,
(b) each P_j in X is false if $j \neq k$,
(c) there is at least one true proposition A that bears the relation R to the ordered pair $<P_k, X>$

where (a) and (b) taken together constitute the *central presupposition*. The canonical form of a *direct answer* to Q is

(*) P_k in contrast to the rest of X because A.

The proposition A is known as the *core of an answer* to Q, because the direct answer would normally be abbreviated, "Because A." The following conditions must be met if (*) is to qualify as a direct answer to Q:

(i) A is true.
(ii) P_k is true.
(iii) No member of X other than P_k is true.
(iv) A bears relation R to $<P_k, X>$.

The context in which a question is posed involves a body of background knowledge K. According to van Fraassen, two of the biggest problems faced by other theories of scientific explanation are *rejections of requests for explanation* and *the asymmetries of explanation*. Rejections are handled in van Fraassen's theory in terms of the presupposition. Unless K entails the truth of the central presupposition—namely, that the topic is true and that every member of the contrast class other than the topic is false—the question does not *arise* in that context. For example, the question "Why was Jimmy Hoffa murdered?" does not arise in my current knowledge situation, for to the best of my knowledge it has not been established that he was murdered—only that he disappeared. In addition, if K entails the falsity of (c)—that is, if K entails that there is no answer—the question does not arise. If, for instance, one asks why a given unstable nucleus decayed at some particular moment, we might answer that there is no reason why. Our best theories tell us that such things just happen. In other words, it is appropriate to raise the question Q if we know that P_k is the one and only member of the contrast class that is true, and we do not know that Q has no answer.

If the question Q does not arise in a given context, we should reject it rather than trying to provide a direct answer. This can be done, as we saw in the preceding paragraph, by providing a corrective answer to the effect that some part of the central presupposition is not entailed by the body of knowledge K, or that K entails that there is no direct answer. If the question does arise, but we find out that (c) is false, then a corrective answer to that effect is appropriate. If the presupposition is completely satisfied, the request for a direct answer is legitimate, and if one is found, it constitutes an explanation of P_k.

There is, I believe, a profound difficulty with van Fraassen's theory centering on the relevance relation R. It can be put very simply, namely, that he imposes no restriction whatever on the nature of the relevance relation. He says explicitly that A is relevant to P_k if A bears relation R to P_k (1980, 143). But if R is not a bona fide relevance relation, then A is 'relevant' to P_k only in a Pickwickian sense. The difficulty can be posed in extremely simple terms. Formally speaking, a relation consists of a set of ordered pairs. Suppose we want to explain any fact P_k. Pick any arbitrary true proposition A. Let the relation R be the unit set of ordered pairs $\{<A, <P_k, X>>\}$ – the set that has $<A, <P_k, X>>$ as its only member. "Because A" is an explanation of P_k.

Although I had studied van Fraassen's theory with some care between 1980 (when it was published) and 1985 (when the Minnesota NEH Institute was held), I had not noticed this difficulty. It emerged as a result of a discussion with Philip Kitcher, and we published a joint paper in which it was exhibited (1987). The problem was masked by a number of van Fraassen's informal remarks. For instance, at the outset of his exposition of his theory of why-questions, he says, "This evaluation [of answers] proceeds with reference to the part of science accepted as 'background theory' in that context" (1980, 141). Earlier, he had remarked that, "To ask that . . . explanations be scientific is only to ask that they rely on scientific theories and experimentation, not on old wives' tales" (1980, 129) and "To sum up: no factor is explanatorily relevant unless it is scientifically relevant; and among the scientifically relevant factors, context determines explanatorily relevant ones" (1980, 126). In conclusion, he says, "To call an explanation scientific, is to say nothing about its form or the sort of information adduced, but only that the explanation draws on science to get this information (at least to some extent) and, more importantly, that the criteria of evaluation of how good an explanation it is, are being applied using a scientific theory" (1980, 155–56). But in the formal account no restriction is imposed on the relation R.

To see the consequences of this lacuna, consider a concrete example. Suppose someone asks why John F. Kennedy died on 22 November 1963; this is the question $Q = <P_k, X, R>$, where

P_k = JFK died 11/22/63   (topic)
X = {JFK died 1/1/63,
     JFK died 1/2/63,

          .

          .

          .

     JFK died 11/22/63,

          .

          .

          .

     JFK died 12/31/63,
     JFK survived 1963}   (contrast class)
R = astral influence   (relevance relation)

Suppose that the direct answer is

P_k in contrast to the rest of X because A,

where A (the core of the answer) consists of a *true* description of the configuration of the planets, sun, moon, and stars at the time of Kennedy's birth. Suppose further that the person who supplies this answer has an astrological theory from which it follows that, given A, it was certain, or highly probable, that Kennedy would die on that day. We now have a why-question and an answer; the answer is an explanation. We must ask how good it is.

Van Fraassen's theory does not stop at the definition of an answer to a why-question; obviously, it must offer grounds for evaluating answers. We need to be able to grade explanations as better or worse. And he does provide criteria for this purpose; there are three. First, we must ask how probable the answer is in light of our background knowledge K. Second, we must ask to what extent the answer *favors* the topic vis-à-vis the other members of the contrast class. Third, we must ask how this answer compares with other available answers: (i) are any other answers more probable? (ii) do any other answers more strongly favor the topic? or (iii) do any other answers render this one wholly or partially irrelevant? On each of these criteria the astrological explanation gets highest marks. In the first place, since we have excellent astronomical records we have practical certainty regarding the celestial configuration at the time of Kennedy's birth; indeed, we have stipulated that A is true. In the second place, we must suppose that the astrologer can derive from A, by means of astrological theory, that Kennedy was sure to die on that day — or, at least, that the probability for his death on that day was much greater than for any other day of the year, and also much greater than the probability that he would live to see the beginning of 1964. Since this explanation, like any explanation, is given ex post facto, we must credit the astrologer with sufficient ingenuity to produce such a derivation. In the third place, no other answer is better than A: (i) A is true; hence, no other answer could be more probable. Moreover, since, astrologically speaking, the heavenly configuration at the time of one's birth is the primary determinant of one's fate (ii) no other answer could favor the topic more strongly and (iii) no other facts could supersede answer A or render it irrelevant.

Consideration of the astrology example makes vivid, I hope, the fundamental problem van Fraassen's theory encounters with respect to the relevance relation. In our critique of that theory, Kitcher and I show formally that *any* true proposition A can be an indispensable part of an explanation of *any* topic P_k (with respect to a contrast class that contains P_k and *any* assortment of false propositions), and, indeed, that it gets highest marks as an explanation of P_k (1987, 319–22). Thus, it is natural to suggest that van Fraassen add one item to the list of presuppositions of Q, namely,

(d) R is a relevance relation.

But when we attempt to impose such a condition, we find ourselves in a mare's nest of difficulties. The problem is to characterize in a general way what constitutes a suitable relevance relation, recognizing, of course, that there may be more than one.

Consider, for example, the relation of logical deducibility. That this relation is not satisfactory has been known since antiquity; Aristotle pointed out that some demonstrations provide understanding while others do not. Hempel recognized from the outset that demonstrations like

Horace is a member of the Greenbury School Board.
All members of the Greenbury School Board are bald.

Horace is bald.

cannot qualify as a bona fide explanation of Horace's baldness. Since the beginning of the first decade in 1948 he has insisted that the demonstration must contain essentially at least one law-statement among its premises. But we have seen what a vexed question it is to distinguish lawlike from nonlawlike statements.

We have also seen that, even if the problem of lawlikeness can be handled, a problem about asymmetries arises. Recalling one of our standard counterexamples, we have two valid deductive arguments with suitable lawful premises, one of which seems clearly to provide an explanation, the other of which seems to most of us not to. From the height of the flagpole in conjunction with the elevation of the sun and laws of propagation of light, we can deduce the length of the shadow it casts. This sounds like a good explanation. From the length of the shadow and other premises of the aforementioned sort (excluding, of course, the height of the flagpole), we can deduce the height of the flagpole. This does not sound like a good explanation. Hence, the problem of asymmetries.

Van Fraassen has maintained, as we have already noted, that the two chief problems inherited from the traditional accounts of explanation that he wants to solve are the problem of rejections and the problem of asymmetries. We have already discussed his solution to the problem of rejection of the explanatory questions, and found no fault with it. His solution to the problem of asymmetries is another story — indeed, it is the fable of "The Tower and the Shadow" (1980, 132–34).

Van Fraassen's treatment of the asymmetries is to show that certain why-questions usually arise in certain typical contexts in which a standard sort of answer is satisfactory. According to van Fraassen's story, when first he asks why the shadow of the tower is so long, he is told that it is cast by a tower of a certain height; in addition, his host, the Chevalier, adds that the tower was built to that height on that particular spot for certain historical reasons. That is *his* explanation, but later in the tale we learn that it is false. The servants have a different, and more accurate, explanation of the position and height of the tower. Carefully taking the contextual factors into account, we discover that the correct answer to the question, "Why is the tower that tall and located in that place," is that it casts a shadow long enough to cover a certain spot on the terrace, where the Chevalier had murdered a servant girl in a fit of jealous rage, at a certain time of day. Here we have an admissible why-question and a suitable direct answer. We are tempted to remonstrate, as I did (1984, 95), that it was the antecedent desire of the Chevalier to have a shadow that long that explains the height of the tower. But van Fraassen seems to be maintaining that the answer given is legitimate. The topic P_k of the question is that the tower stands at a particular place and is 175 feet tall. The contrast class X consists of a series of statements about towers of different heights located at various locations in the vicinity. And the relation R is a relation of intentional relevance. Our remonstrance was based on the belief that this relation is not a suitable explanatory relevance relation. As an advocate of a causal/mechanical conception of scientific explanation, I am not prepared to admit that effects explain their causes, even where conscious purposeful behavior is involved.

It is clear from the whole tenor of van Fraassen's account that he is not proposing an 'anything goes' sort of theory. He is *not* suggesting that any answer to a why-question that happens to satisfy a questioner is a satisfactory explanation. Consequently, it seemed to Kitcher and me, van Fraassen's pragmatic theory cannot escape precisely the kinds of problems concerning objective explanatory relevance relations with which other more traditional theories — such as the received view or the statistical-relevance approach — had to struggle. Hence, when I proposed above the addition of presupposition

(d) R is a relevance relation

I was opening up the whole question of what constitutes a satisfactory explanatory relevance relation.

Consider, for example, the relation of statistical relevance, which I once regarded as the key explanatory relation. We suppose that the person who raises a given why-question has a prior probability distribution over the members of the contrast class. When an appropriate answer is given, it results in a different probability distribution over the same set of alternatives. In the S-R model I required only that the probabilities change; van Fraassen's theory requires that the topic be favored—that is, that it be elevated relative to its rivals in the contrast class. What kinds of probabilities are these? Coffa and I took propensities and frequencies, respectively, as the appropriate interpretations. In a pragmatic theory of explanation, because of its emphasis upon the knowledge situation and the context, it is natural to think of epistemic or personal probabilities. Given the well-known difficulties with the former, van Fraassen seems to prefer the latter.

Whatever sort of probability is involved, a basic problem—one we have confronted previously—arises. When an explanation is sought we already know that the explanandum-phenomenon has occurred. For van Fraassen, the truth of the topic is the first presupposition of a why-question. Moreover, the second presupposition of the why-question is that all of the other members of the contrast class are false. The why-question does not arise unless these presuppositions are fulfilled. But if this presupposition is fulfilled, the prior probability distribution with respect to the body of knowledge K is one for the topic and zero for all other members of the contrast class. No change in this probability distribution could possibly favor the topic. So van Fraassen proposes that we must cut back our body of knowledge K to some proper part K(Q) that is appropriate for question Q (1980, 147). The problem of deciding what information K(Q) should contain is precisely the problem Hempel faced in connection with the requirement of maximal specificity. It is also the problem with which I had to deal in characterizing objective homogeneity. It turns out, then, that this cluster of traditional problems is not evaded by van Fraassen's pragmatic account.

The appeal to personal probabilities in this context gives rise to another serious difficulty. Return to the example of John F. Kennedy's assassination. To the sincere believer in astrology the configuration of heavenly bodies at the time of Kennedy's birth is highly relevant to his death on that particular fateful day in 1963. Acquiring that information will produce a redistribution of personal probabilities strongly favoring the topic. Believing, as we do, that there is no objective relevance between the celestial configuration at the time of Kennedy's birth and the occurrence of his death on a particular day, we need to block such explanations. Unless we can impose the demand for objective relevance relations, we cannot arrive at a satisfactory characterization of scientific explanation. As many philosophers have insisted, we need to appeal to objective nomic relations, causal relations, or other sorts of physical mechanisms if we are to provide adequate scientific explanations.

Philosophers have long recognized that scientific explanation has pragmatic dimensions. It is obvious that various features of actual explanations depend upon context. Hempel was aware from the beginning that the individual who answers an explanation-seeking why-question standardly omits parts of the explanation that are already well known to the questioner. I have emphasized the fact that one frequently has to clarify the question in order to ascertain what explanation is being sought. As we saw in the second decade, the ordinary language philosophers placed great emphasis upon pragmatic considerations. The theory of the pragmatics of explanation given by van Fraassen in *The Scientific Image* is highly illuminating and is, I believe, the best that has been given to date. It must be emphasized, however, that he has not succeeded in showing that all the traditional problems of explanation can be solved by appealing to pragmatics. In that sense he has *not* provided a pragmatic theory of explanation. The problems concerning the nature of laws, and those concerning the nature of causality, have not been circumvented by pragmatic considerations.

Another important representative of the erotetic version of the epistemic conception is Peter Achinstein, whose view is articulated in great detail in *The Nature of Explanation* (1983). His theory differs significantly from those of both Bromberger and van Fraassen, but it is much closer in spirit to that of Bromberger. This can be seen most clearly in the emphasis that both Achinstein and Bromberger place on the linguistic analysis of English usage — something van Fraassen does hardly at all. Whereas van Fraassen simply announces that explanations are answers to why-questions, Achinstein invests considerable time and effort in clarifying the usage of "explanation" and closely related terms. Along with Bromberger, he denies that all explanations are answers to why-questions. Nevertheless, as we shall see, questions — not just why-questions — and their answers play a fundamental role in this theory.

Achinstein points out that "explanation" may refer either to a *process or a product*. The process is a linguistic performance; someone explains something to someone by uttering or writing statements. The product is the content of the linguistic performance. The linguistic performance itself involves an intention on the part of person producing the explanation: "Explaining is what Austin calls an illocutionary act. Like warning and promising, it is typically performed by uttering words in certain contexts with appropriate intentions" (1983, 16). According to Achinstein, this process concept of explanation is primary. When we try to characterize the product, we must take account of the intention (or illocutionary force) of the explanation, for the same set of words can be used either to explain or do other sorts of things. A physician might explain John's malaise this morning by saying, "He drank too much last night." John's wife might use the same words to criticize his behavior. His wife's speech act is not an explanation; consequently, what she produced is not an explanation. This same pragmatic consideration arose, incidentally, in the 1948 Hempel-Oppenheim essay for, given the explanation/prediction symmetry thesis, an argument may function in one context as an explanation and in another as a prediction. To deal explicitly with this aspect of explanation, Achinstein adopts what he calls *an ordered pair view* of explanation: an explanation in the product sense is an ordered pair < x, y > in which x is a specified type of proposition and y is a type of speech act, namely, explaining. On this view, y retains the intention involved in the process of explanation (1983, 85–94). Achinstein refers to his account as *the illocutionary theory*.

It should be noted that the foregoing considerations are designed to clarify the notion of explanation without qualification. Up to this point, Achinstein makes no attempt to characterize correct explanations, good explanations, or scientific explanations. It is not until page 117, almost one-third of the way through a fairly large book, that scientific explanation comes up for serious consideration. The philosopher of science who is impatient to get to that topic cannot begin there, however, for the preliminaries are used extensively in the subsequent discussions.

In his preliminary formulations, Achinstein presents two aspects of explanation:

(1) If S explains q by uttering u, then S utters u with the intention that his utterance of u render q understandable (1983, 16);

(2) If S explains q by uttering u, then S believes that u expresses a proposition that is a correct answer to Q (1983, 17).

If Q is a why-question (which is, I repeat, for Achinstein only one among many sorts of explanation-seeking questions), then q is what van Fraassen called the *topic* of the question. To use one of Achinstein's examples, if Q is "Why did Nero fiddle?" then q is "Nero fiddled." According to (1), whatever answer is given is intended to make q understandable, but since explanation and understanding are such closely related concepts, it is necessary for Achinstein to say something about what constitutes understanding. He offers the following necessary condition:

A understands q only if there exists a proposition p such that A knows of p that it is a correct answer to Q, and p is a complete content-giving proposition with respect to Q. (Here p is a proposition expressed by a sentence u uttered by A.) (1983, 42)

He later suggests that it is a sufficient condition as well (1983, 57). Space does not permit a full statement of what constitutes a complete content-giving proposition with respect to a question; the details are spelled out in Achinstein's book. But the crucial point can be raised by considering a special case — one of his examples. A straightforward complete content-giving proposition with respect to the question, "Why did Nero fiddle?" is "The reason Nero fiddled is that he was happy." Given that this is a complete content-giving proposition with respect to that question, A understands Nero's fiddling iff A knows that "The reason Nero fiddled is that he was happy" is a *correct answer* to the question, "Why did Nero fiddle?"

This view of explanation seems seriously question-begging. We may raise essentially the same question with regard to Achinstein's theory as we did concerning van Fraassen's: what objective relationship must obtain between the fact that Nero was happy and the fact that he fiddled to make "The reason Nero fiddled is that he was happy" a correct answer? How must that relationship differ from the relationship between the fact that Caesar was assassinated on the Ides of March and the fact that Nero fiddled? These questions have fundamental importance; to see this it will be useful to make a direct comparison with Hempel's theory of deductive-nomological explanation.

According to Hempel's theory, if "Nero was happy" is part of the explanans of "Nero fiddled," then "Nero was happy" must be true. Hempel's theory of explanation specified the relationship that must exist between the facts if one is to be (part of) an explanation of the other. According to the D-N model, the statement "Nero was happy" must be a premise of a valid deductive argument having "Nero fiddled" as its conclusion, and including essentially at least one other premise stating a lawful regularity. This argument must fulfill the Hempel-Oppenheim empirical condition of adequacy, namely, that all of its premises be true. Given the fulfillment of these conditions, we are then authorized to accept the claim that (at least part of) the reason Nero fiddled is that he was happy. It was *not* part of the empirical condition of adequacy to determine that "The reason Nero fiddled is that he was happy" is true. The whole idea of the Hempel-Oppenheim theory was to provide conditions under which it is correct to make such claims as "The reason Nero fiddled is that he was happy." They do not require us to assess the truth of such a statement to ascertain whether or not we have a correct explanation. Achinstein is clearly aware that this line of argument may be brought against his theory, and he attempts to rebut it (1983, 71–72), but I am not convinced that his defense is successful. I think he ends up—like Bromberger and van Fraassen—lacking an adequate characterization of the kinds of objective relevance relations required in sound scientific explanations.

Despite my skepticism regarding the illocutionary theory of explanation, there are, it seems to me, several especially illuminating features of Achinstein's treatment of scientific explanation. First, he distinguishes carefully between *correct* explanations and *good* explanations. An explanation < x, y > is a correct explanation if the first member of that ordered pair is a true statement. Of course, the fact that the ordered pair is an explanation imposes other qualifications on x. However, for any number of pragmatic reasons, a correct explanation may not be a good explanation. It may be unsuitable to the knowledge and abilities of the listeners, or lacking in salience with respect to their interests. To deal with evaluations of explanations over and above correctness, Achinstein introduces the idea of a set of instructions to be followed in constructing explanations (1983, 53–56).

Such instructions could be of a wide variety of kinds. One might be the Hempelian instruction that the explanation must include essentially at least one law. Another might be an instruction to give a microphysical explanation. Still another might be the instruction to give a causal explanation.

Another important feature is Achinstein's contention that there is no single set of universal instructions that will suffice to judge the merits of scientific explanations at all times and in all contexts. Instructions that are suitable in one context may be quite unsuitable in another. He offers a characterization of appropriate instructions as follows:

> I is a set of appropriate instructions for an explainer to follow in explaining q to an audience iff *either*
> a. The audience does not understand q in a way that satisfies I, and
> b. There is answer to Q (the question that elicits an explanation of q), that satisfies I, the citing of which will enable the audience to understand q in a way that satisfies I, and
> c. The audience is interested in understanding q in a way that satisfies I, and
> d. Understanding q in a way that satisfies I, if it could be achieved, would be valuable for the audience;

or

> It is reasonable for the explainer to believe that a-d are satisfied. (113, slightly paraphrased)

Employing several important historical examples, Achinstein argues that there is no set of instructions that is universally appropriate for science (1983, 119–56).

The discussion of universal instructions leads naturally into another main feature of Achinstein's theory, namely, a consideration of the possibility of formal models of explanation such as the D-N, I-S, S-R. Again the conclusion is negative. Achinstein states two requirements which, he believes, motivate the "modelists." The first is the "No-Entailment-By-Singular-Sentence" (or NES) requirement. According to this requirement, no correct explanation of a particular occurrence can contain, in the explanans, any singular sentence or finite set of singular sentences that entail the explanandum (1983, 159). The second requirement he calls "the a priori requirement." According to this requirement, "the only *empirical* consideration in determining whether the explanans correctly explains the explanandum is the truth of the explanans; all other considerations are a priori" (1983, 162). His strategy in arguing that there cannot be models of explanation, in the traditional sense, is that any model that satisfies one of these requirements will violate the other (1983, 164–92). Even though I have spent a great deal of effort in elaborating the S-R model, this view of Achinstein's is one with which I happen to agree. I now think that an adequate scientific explanation identifies the mechanisms by which the explanandum came about. Consequently, what constitutes a suitable scientific explanation depends on the kinds of mechanisms—

causal or noncausal – that are operative in our world. This is an issue that cannot be settled a priori. Achinstein's arguments against the possibilities of models of explanation are far more precise and detailed than mine.

4.5 Empiricism and Realism

Van Fraassen's seminal book, *The Scientific Image*, has spawned a great deal of discussion, most of which has been directed toward his rejection of scientific realism. Churchland and Hooker, *Images of Science* (1985), is an important collection of critical essays, accompanied by van Fraassen's detailed replies. None of these essays focuses primarily on his treatment of scientific explanation, though many are tangentially relevant to it.[21] Since, however, the realism issue has direct bearing on the nature of theoretical explanation, we must devote some attention to it. To carry out this discussion I will accept van Fraassen's claim – which I believe to be sound – that there is a viable distinction between the observable and the unobservable. Although the dividing line may not be sharp, there are clear cases of observables (e.g., sticks and stones) and clear cases of unobservables (e.g., atoms, electrons, and simple molecules).

The first sentence in the first chapter of *The Scientific Image* is striking: "The opposition between empiricism and realism is old, and can be introduced by illustrations from many episodes in the history of philosophy" (1980, 1). It formulates an assumption that goes unquestioned throughout the rest of the book – namely, that it is impossible to have empirical evidence that supports or undermines statements about objects, events, and properties that are not directly observable by the unaided normal human senses. This assumption should not, I think, go completely unchallenged. Indeed, I believe it is false. However that may be, it raises what I take to be the *key question* for scientific empiricism (W. Salmon 1985).

Let me illustrate the point by means of a simple example. I own a copy of the Compact Edition of the Oxford English Dictionary. It contains print of various sizes. I can read the largest print on the title page with my naked eye, but everything else is blurry without my eyeglasses. When I put on the spectacles I can easily read some of the larger print within the books. The use of corrective lenses does not take us beyond the realm of the directly observable; their effect is to restore normal vision, not to extend it beyond the normal range. The spectacles enable me to see the things I could have seen in my early teens without their aid. But even with their aid much of the print is blurry. With the aid of a magnifying glass (which comes with the set) I can read even the entries in smallest type. Also, with the aid of the magnifying glass, I can see marks of punctuation that were completely invisible to me without it. I claim that I have established the existence of an entity that is not directly observable; I have established a statement to the effect that there is, at a particular place on a given page, an ink spot too small to be detected with the normal unaided human sense of sight. Moreover, I cannot feel it, smell it, hear it, or taste it.

It is important to note that, when I view through the magnifying glass print that I *can* read without it, the letters, words, and marks of punctuation that I see are the same. They simply appear clearer and larger. When I view smaller print with the magnifying glass, the forms I see—letters, words, marks of punctuation—make sense. I see bona fide words, and I read appropriate definitions of them. The words I read appear in their correct places. The same is true of the dot that I could not see at all without the magnifying glass. When it is made visible by the glass, it appears in a syntactically correct place. Moreover, although I confess that I have not performed the experiment, I have complete confidence that a comparison of the entries in the compact edition with those of the unreduced editions would reveal an identity between what is seen without the magnifying glass in the larger edition with what is seen with the aid of the magnifying glass in the compact edition.

Evidently, many different experiments of the type just described can be conducted with a variety of lenses, and on their basis we can establish a theory of geometrical optics. A number of fundamental facts of geometrical optics were known in antiquity and medieval times, and Snell's law was proposed in 1621. This theory is completely empirical even in the narrow sense van Fraassen adopts. With geometrical optics we can develop the theories of the telescope and the microscope. These theories are readily confirmable by means of experiment. Telescopes, for example, can be used to view from a distance terrestrial objects that can be approached for viewing at close range. It is interesting that van Fraassen regards as observable celestial objects—such as the moons of Jupiter—that can be seen from earth only with the aid of a telescope. We can, in principle, travel closer to them and see them with the naked eye. More interesting still is the fact that he takes objects remote from us in time—such as dinosaurs—to be observables. Time-travel into the past is something NASA has not achieved even in a small way. In contrast, he considers objects that can be seen only with the aid of a microscope as unobservables. There is no other place to go to get a better vantage point.

To substantiate my own claim that we can have empirical knowledge of unobservables, I attempted to spell out the sort of inference that underlies the transition from observation to conclusions about unobservables (1984, 231–34). Shortly thereafter I put the matter in these terms:

> When David Hume stated that all of our reasonings about unobserved matters of fact are based upon the relation of cause and effect he was, I suspect, almost completely correct. One notable exception is induction by simple enumeration (or some other very primitive form of induction). As remarked above, I am assuming for purposes of this discussion that some sort of primi-

tive induction is available; I shall not attempt to characterize it or justify it in this context. The type of argument required in connection with microscopic observation is, I think, causal. It is a rather special variety of causal inference that is also analogical. This particular sort of causal/analogical argument is, in my view, quite powerful. To employ it, we must assume that we already have knowledge of cause-effect relations among observables — e.g., that hitting one's thumb with a hammer causes pain, that drinking water quenches thirst, and that flipping a switch turns off the light. Such relations can be established by Mill's methods and controlled experiments. Neither the instrumentalist's nor the constructive empiricist's account of science can get along without admitting knowledge of such relations. Using relations of this sort, we can now schematize what I suspect is the basic argument enabling us to bridge the gap between the observable and the unobservable. It goes something like this:

It is observed that:

An effect of type E_1 is produced by a cause of type C_1.
An effect of type E_2 is produced by a cause of type C_2.

$\cdot$

$\cdot$

$\cdot$

An effect of type E_k occurred.

We conclude (inductively) that:

A cause of type C_k produced this effect of type E_k.

The particular application of this argument that interests us is the case in which $C_1, C_2, \ldots, C_k$ are similar in most respects except size. Under these circumstances we conclude that they are similar in causal efficacy. (1985, 10–11)

The foregoing argument connects quite directly with an analysis of microscopic observation presented by Ian Hacking just after the publication of van Fraassen's *The Scientific Image*. To learn something about the scientific use of microscopes Hacking did something quite extraordinary for a philosopher. He actually went to a laboratory where the use of microscopes is essential to the research being conducted, and he learned how to use a variety of types. The research involved observation of dense bodies in red blood cells, and it employed a standard device known as a *microscopic grid*. "Slices of a red blood cell are fixed upon a microscopic grid. This is literally a grid: when seen through a microscope one sees a grid each of whose squares is labelled with a capital letter" (1981, 315). Making reference to the microscopic grid, he then addresses the issues raised by van Fraassen:

I now venture a philosopher's aside on the topic of scientific realism. Van Fraassen says we can see through a telescope because although we need the telescope to see the moons of Jupiter when we are positioned on earth, we could go out there and look at the moons with the naked eye. Perhaps that fantasy is close to fulfillment, but it is still science fiction. The microscopist avoids fantasy. Instead of flying to Jupiter he shrinks the visible world. Consider the grid that we used for re-identifying dense bodies. The tiny grids are made of metal: they are barely visible to the naked eye. They are made by drawing a very large grid with pen and ink. Letters are neatly inscribed by a draftsman at the corner of each square on the grid. Then the grid is reduced photographically. Using what are now standard techniques, metal is deposited on the resulting micrograph. . . . The procedures for making such grids are entirely well understood, and as reliable as any other high quality mass production system.

In short, rather than disporting ourselves to Jupiter in an imaginary space ship, we are routinely shrinking a grid. Then we look at the tiny disk and see exactly the same shapes and letters as were drawn in the large by the first draftsman. It is impossible seriously to entertain the thought that the minute disk, which I am holding by a pair of tweezers, does not in fact have the structure of a labelled grid. I know that what I see through the microscope is veridical because we *made* the grid to be just that way. I know that the process of manufacture is reliable, because we can check the results with any kind of microscope, using any of a dozen unrelated physical processes to produce an image. Can we entertain the possibility that, all the same, this is some kind of gigantic coincidence[?] Is it false that the disk is, in fine, in the shape of a labelled grid? Is it a gigantic conspiracy of 13 totally unrelated physical processes that the large scale grid was shrunk into some non-grid which when viewed using 12 different kinds of microscopes still looks like a grid? (1981, 316–17)

To avoid possible misinterpretation I should report that Hacking does not mean to use his argument to support the kind of wholesale realism I argued for in "Why Ask, 'Why?'?" but he does offer a strong argument for the conclusion that we can, with the aid of microscopes, have knowledge of objects and properties not visible to the naked eye.[22]

It seems to me that we can usefully distinguish between direct and indirect observation, where direct observation is accomplished by the use of unaided normal human senses, and indirect observation is accomplished by the use of instruments, such as the microscope and the telescope, that extend the range of the senses. I consider my simple argument, based in part on the development of geometrical optics, and Hacking's argument, based on sophisticated microscopy, strong arguments to the effect that we can have indirect observational knowledge of objects and properties that are not directly observable. In addition, I want to claim, our knowledge of unobservables can be extended even further by appealing to appropriate theoretical considerations. That was the upshot of the discussion of Perrin's argument concerning the ascertainment of Avogadro's number N and the issue of molecular reality in §4.2 above. Perrin studied the behavior of Brownian particles that could be observed using microscopes available in his day. He used his *indirect observations* of these entities as a basis for *inferring* the existence of simple molecules that are much too small to be viewed microscopically. It is mildly amusing that Perrin cited 13 independent methods of ascertaining N and that Hacking refers to 13 independent physical processes in dealing with the microscopic grid. The number 13 appears to be especially unlucky for antirealists.

One problem that has traditionally been associated with scientific realism is the problem of the meaning of theoretical terms. That cannot be the crucial problem, for we have been entertaining claims about unobservable entities without using any esoteric theoretical vocabulary. Since we do successfully describe the things we directly observe, our language must contain an observational vocabulary. However we might characterize it in general, "ink spot," "page," and "smaller than" are surely terms within it. Recall, in this connection, Perrin's work with tiny spheres of gamboge — *much* too small to be observed directly — which can be described entirely within our observational vocabulary.

I realize, of course, that there exists a broad range of philosophical opinion on the matter of scientific realism. I have recapitulated the argument that I find compelling. Other philosophers — e.g., Arthur Fine and Larry Laudan — join van Fraassen in rejecting realism, but for reasons quite different from van Fraassen's. Still others — e.g., Richard Boyd and Ernan McMullin — embrace realism, but appeal to different kinds of arguments to support their views.[23] We cannot escape the realism issue, for it is crucial to the debate between Coffa's highly realistic ontic conception of scientific explanation and van Fraassen's highly pragmatic erotetic approach.

4.6 Railton's Nomothetic/Mechanistic Account

In the first year of the fourth decade Peter Railton's first published article on scientific explanation appeared (1978). It is addressed chiefly to Hempel's I-S model, and it embodies an attempt to provide an account of probabilistic explanation that avoids what Railton regards as the two most troublesome aspects of Hempel's model — namely, epistemic relativization and the requirement of maximal specificity. This article can be viewed, in part, as a further development of some of the ideas expressed in Jeffrey (1969). With Jeffrey, Railton rejects the thesis of the received view that all explanations are arguments, but he goes further than Jeffrey in this regard. Whereas Jeffrey had admitted that in some "beautiful cases" (where the difference between the actual probability and unity is so small as to "make no odds") statistical explanations can be arguments, Railton argues that (practical considerations aside) there is no theoretical difference between the beautiful cases and the unbeautiful. In none of the kinds of cases Hempel treated as inductive-statistical, he claims, should the explanation be construed as an argument. Railton also agrees with Jeffrey in maintaining that, for ineluctably statistical phenomena, the key to explanation lies in understanding the stochastic mechanism by which the occurrence came to pass, not in finding some way to render the event nomically to be expected. In this connection, both Jeffrey and Railton agree that, where some results of a given stochastic process are probable and others improbable, we understand the improbable just as well as we understand those that are highly probable. With Jeffrey, consequently, Railton rejects the high-probability requirement. Railton goes far beyond Jeffrey, however, in spelling out the details of a model of scientific explanation that embodies all of these features.

Railton chooses to elaborate what he calls a *deductive-nomological model of probabilistic explanation* — or D-N-P *model* — in terms of an example of an event that has an extremely small probability, namely, the alpha decay of a nucleus of uranium^{238}. Since the mean-life of this radionuclide is 6.5×10^9 years, the probability that such an atom would decay within a specific short period of time is almost vanishingly small — but not quite vanishing, for such decays do occur. Our theory enables us to calculate the probability p that such a decay will occur within a short time interval Δt. Suppose u, a particular nucleus of this sort, has emitted an alpha-particle during such an interval. Then, we can set up the following deductive argument:

(2) (a) All nuclei of U^{238} have probability p of emitting an alpha-particle during any interval of length Δt, unless subjected to environmental radiation.

(b) Nucleus u was a nucleus of U^{238} at time t and was subjected to no environmental radiation during the interval $[t, t + \Delta t]$.

(c) Nucleus u had a probability p of emitting an alpha-particle during the interval $[t, t + \Delta t]$. (Railton 1978, 214, slightly paraphrased)

In this argument Railton, like Coffa, construes the probabilities as single-case propensities, and he regards premise (a) as a lawful generalization. He assumes, moreover, that (a) is an irreducibly probabilistic law, and that it incorporates all probabilistically relevant factors. He recognizes that (2) appears to be an explanation of the fact that u had a probability p of decaying within the specified time interval; he maintains, however, that it can be supplemented in a way that transforms it into an explanation of the fact that u actually decayed. The first addition is "a derivation of (2a) from our theoretical account of the mechanism at work in alpha-decay" (1978, 214) — from the theory of quantum-mechanical tunneling. This is, I take it, another deductive argument employing the Schrödinger wave equation and such facts as the atomic weight and atomic number of $_{92}U^{238}$ as its premises. It would qualify as one of Hempel's D-S explanations.

Railton's approach is strongly mechanistic, as can be seen clearly by contrasting his attitude toward these two arguments with Hempel's attitude. For Hempel, we recall, to explain a particular fact it is sufficient to subsume it under a law. Such explanations are complete. If one wants an explanation of a law that entered into the first explanation, it can be supplied by deriving that law from more general laws or theories. The result is another explanation. The fact that a second explanation of this sort can be given does nothing to impugn the credentials of the first explanation. Railton's view is quite different. According to him—and in this I heartily agree—explanation involves revealing the mechanisms at work in the world. Mere subsumption of phenomena under generalizations does not constitute explanation. Explanation involves understanding *how the world works*. For Railton, then, the quantum mechanical explanation of the probabilistic decay law is an integral and indispensable part of the explanation of the decay that occurred.

To transform this pair of arguments, which still appear to explain the probability of the decay of u, into an explanation of the actual decay of u, we are asked to supply a "parenthetic addendum to the effect that u did alpha-decay during the interval" in question (1978, 214). These three components, taken together, though they do not constitute an argument or a sequence of arguments, do constitute an *explanatory account*:

    (3) A derivation of (2a) from our theoretical account of the mechanism of alpha decay.
        The D-N inference (2).
        The parenthetic addendum.

A D-N explanation and a D-S explanation (or two D-N explanations, if we continue to consider D-S a subtype of D-N) serve as the core of the explanatory account, but they do not comprise the whole explanation. The parenthetic addendum is also required. If the parenthetic addendum were taken as an additional premise for an argument, the explanation would be vitiated by becoming trivially circular.

If the explanatory account is not an argument, Railton realizes, many readers are going to wonder about its explanatory status:

Still, does (3) explain why the decay took place? It does not explain why the decay *had to* take place, nor does it explain why the decay *could be expected to* take place. And a good thing, too: there is no *had to* or *could be expected to* about decay to explain—it is not only a chance event, but a very improbable one. (3) does explain why the the decay *improbably* took place, which is how it did. (3) accomplishes this by demonstrating that there existed at the time a small but definite physical possibility of decay, and noting that, by chance, this possibility was realized. The derivation of (2a) that begins (3) shows, by assimilating alpha-decay to the chance process of potential barrier tunneling, how this possibility comes to exist. If alpha-decays are chance phenomena of the sort described, then once our theory has achieved all that (3) involves, it has explained them to the hilt, however unsettling this may be to *a priori* intuitions. To insist upon stricter subsumption of the explanandum is not merely to demand what (alas) cannot be, but what decidedly should not be: sufficient reason that one probability rather than another be realized, that is, chances without chance. (1978, 216)

Railton argues carefully that his characterization of probabilistic explanation does escape epistemic relativization, and that it has no need for a requirement of maximal specificity. The basic reason is that, by construing (2a) as a (putative) law, he claims that it is simply false if it fails to be maximally specific. If there are further factors relevant to the occurrence of the decay, which have not been included in the alleged law, it is false. Moreover, the kind of maximal specificity to which he refers is fully objective; it is just as objective as my objectively homogeneous reference classes. Inasmuch as Railton adopts a *single case propensity interpretation* of probability, he has no need for reference classes, but the maximal specificity of his laws is strictly analogous to objective homogeneity for a frequentist (see the discussion of propensities in §3.3 above).

In the introductory sections of his 1978 paper, Railton offers some extremely compact remarks about his approach to explanation in general, not just probabilistic explanations of particular facts. Regarding his mechanistic orientation, he says,

The goal of understanding the world is a theoretical goal, and if the world is a machine — a vast arrangement of nomic connections — then our theory ought to give us some insight into the structure and workings of the mechanism, above and beyond the capability of predicting and controlling its outcomes. . . . Knowing enough to subsume an event under the right kind of laws is not, therefore, tantamount to knowing the *how* and *why* of it. As the explanatory inadequacies of successful practical disciplines remind us: explanation must be more than potentially-predictive inferences or law-invoking recipes. (1978, 208)

Some of the mechanisms are, of course, indeterministic.

The D-N probabilistic explanations to be given below do not explain by giving a deductive argument terminating in the explanandum, for it will be a matter of chance, resisting all but *ex post facto* demonstration. Rather, these explanations subsume a fact in the sense of giving a D-N account of the chance mechanism responsible for it, and showing that our theory implies the existence of some physical possibility, however small, that this mechanism will produce the explanandum in the circumstances given. I hope the remarks just made about the importance of revealing mechanisms have eased the way for an account of probabilistic explanation that focuses on the indeterministic mechanisms at work, rather than the "nomic expectability" of the explanandum. (1978, 209)

The views expressed by Railton in this early article are ones I find highly congenial. There may, however, be a difference in our attitudes toward causal explanation. Like Jeffrey before him, Railton discusses the old chestnut of the falling barometric reading and the storm:

> (S) The glass is falling.
>   Whenever the glass falls the weather turns bad.
>   _____
>   The weather will turn bad.

and considers repairing it by adding a causal premise:

> (C) The glass is falling.
>   Whenever the glass is falling the atmospheric pressure is falling.
>   Whenever the atmospheric pressure is falling the weather turns bad.
>   _____
>   The weather will turn bad.

Now Railton considers (C) a causal explanation and its third premise a causal law. He points out that, until we understand the mechanism behind that causal relationship, we do not have an adequate explanation of the turn in the weather. He therefore concludes that we need something over and above the causal infusion that transformed (S) into (C). I agree that we need something more, but we also need something less.

Paul Humphreys has noted (in a personal communication) that there is something odd about the second premise of (C) as a component in a causal explanation, for its presence licenses an inference from effect to cause. And once that point is clear, we see that the first premise of that argument has no place in a causal explanation (or any other, I should think). As we noted in our earlier analysis of this example, the storm and the falling barometer are effects of a common cause, the drop in atmospheric pressure, and the cause screens the common effects off from one another. Therefore (C) should be replaced by

> (C') The atmospheric pressure is falling.
>   Whenever the atmospheric pressure is falling the weather turns bad.
>   _____
>   The weather will turn bad.

The first premise of (C) is no part of the explanation; it is our evidence for the truth of the first premise of (C').

Still, I would not call (C') causal, for my conception of causality includes mechanisms of propagation and interaction. But I would regard the explanation that results from supplementation of (C')—in the way we both consider appropriate—a causal explanation. So far, then, the main disagreement is terminological. But Railton also makes mention of such structural laws as the Pauli exclusion principle, which he rightly holds to be noncausal (1978, 207). He characterizes explanations based upon such laws as noncausal *structural explanations*. In *Scientific Explanation and the Causal Structure of the World* I had taken the attitude that structural laws have explanatory force only if they themselves can be explained causally. In that context I was thinking of such structural laws as the ideal gas law, and I placed great weight upon its explanation on the basis of kinetic theory. Since I did not endeavor to offer an account of quantum mechanical explanation, I did not confront such laws as the Pauli exclusion principle.[24]

At the time Railton sent this article off for publication he must have been working on his dissertation (1980). The brief suggestive remarks he offers in the article, in addition to his articulation of the D-N-P model, are elaborated clearly and at length in the dissertation—a monumental two-volume work that comes to 851 pages in toto.

When first I read the 1978 essay, I must confess, I failed to understand it. Only after seeing a later paper did I begin to appreciate the beauty of Railton's work on explanation. In 1981 a landmark symposium on probabilistic explanation (Fetzer 1981) was published; it included, among many other valuable contributions, Railton's "Probability, Explanation, and Information." In this essay he offers further discussion of the D-N-P model, addressing certain objections that might be brought against it. Of particular importance is the objection based on the acknowledged fact that many proffered explanations that are widely accepted as correct omit items, such as laws, that form an integral part of any explanation that conforms to the D-N-P model.

To deal with this problem, Railton introduces a distinction—one that turns out to be extraordinarily fruitful—between an *ideal explanatory text* and *explanatory information* (1981, 240). Given an event we wish to explain, the ideal explanatory text would spell out all of the causal and nomic connections that are relevant to its occurrence. For probabilistic explanation, the D-N-P model furnishes the schema for the ideal explanatory text. The ideal explanatory text can be expected—in most, if not all, cases—to be brutally large and complicated. When one considers the myriad molecules, atoms, subatomic particles, and interactions involved in everyday events, it is easy to see that the ideal explanatory text is an *ideal* that might never be realized. That does not matter.[25] The scientist, in seeking scientific understanding of aspects of our world, is searching for *explanatory information* that enables us to fill out parts of the ideal explanatory text. The ideal explanatory text constitutes a framework that provides guidance for those who endeavor to achieve understanding of various aspects of the world, and that is its primary function.

Railton acknowledges the possibility that, in addition to the probabilistic explanations characterized by the D-N-P model, there may be nonprobabilistic explanations that are closely related to Hempel's D-N model. For Railton, however, D-N explanations are far more robust than they are for Hempel:

> I would argue that the D-N schema instead provides the skeletal form for ideal explanatory texts of non-probabilistic phenomena, where these ideal texts in turn afford a yardstick against which to measure the explanatoriness of proffered explanations in precisely the same way that ideal D-N-P texts afford a yardstick for proffered explanations of chance phenomena. Thus, proffered explanations of non-probabilistic phenomena may take various forms and still be successful in virtue of communicating information about the relevant ideal text. For example, an ideal text for the explanation of the outcome of a causal process would look something like this: an inter-connected series of law-based accounts of all the nodes and links in the causal network culminating in the explanandum, complete with a fully detailed description of the causal mechanisms involved and theoretical derivations of all of the covering laws involved. This full-blown causal account would extend, via various relations of reduction and supervenience, to all levels of analysis, i.e., the ideal text would be closed under relations of causal dependence, reduction, and supervenience. It would be the whole story concerning why the explanandum occurred, relative to a correct theory of the lawful dependencies of the world. (1981, 246–47)

Does the conception of an ideal explanatory text have any utility? Railton replies,

> [Is it] preposterous to suggest that any such ideal could exist for scientific explanation and understanding? Has anyone ever attempted or even wanted to construct an ideal causal or probabilistic text? It is not preposterous if we recognize that the actual ideal is not to *produce* such texts, but to have the ability (in principle) to produce arbitrary parts of them. It is thus irrelevant whether individual scientists ever set out to fill in ideal texts as wholes, since within the division of labor among scientists it is possible to find someone (or, more precisely, some group) interested in developing the ability to fill in virtually any aspect of ideal texts—macro or micro, fundamental or "phenomenological," stretching over experimental or historical or geological or cosmological time. A chemist may be uninterested in how the reagents he handles came into being; a cosmologist may be interested in just that; a geologist may be interested in how those substances came to be distributed over the surface of the earth; an evolutionary biologist may be interested in how chemists (and the rest of us) came into being; an anthropologist or historian may be interested in how

man and material came into contact with one another. To the extent that there are links and nodes, at whatever level of analysis, which we could not even in principle fill in, we may say that we do not completely understand the phenomenon under study. (1981, 247–48)[26]

The distinction between the ideal explanatory text and explanatory information can go a long way, I think, in reconciling the views of the pragmatists and the realists.[27] In his work on deductive-nomological explanation, Hempel steadfastly maintained an attitude of objectivity. D-N explanations were never epistemically relativized, and they always fulfill the requirement of maximal specificity trivially. As we have seen, Hempel found it necessary to relinquish this attitude when he confronted inductive-statistical explanations, but Coffa, Railton, and I, among others, made serious efforts to restore full objectivity to the domain of probabilistic explanation. At the same time, those philosophers who have emphasized the pragmatics of explanation, from Hanson and Scriven to van Fraassen, have criticized objectivist accounts for demanding inclusion of too much material to achieve what they regard as legitimate explanations. This opposition has, for example, led to the rejection of the covering law conception of explanation by pragmatists, and to insistence upon it by the objectivists.

The issue of how 'fat' or 'thin' explanations should be was raised in *Scientific Explanation and the Causal Structure of the World* (1984, 131–34). and it resurfaced in the 1985 American Philosophical Association Symposium on that book (Kitcher, van Fraassen, and W. Salmon, 1985). One useful way to think about this conflict, I believe, is to regard the objectivists—the advocates of the ontic conception—as focusing on the ideal explanatory text. We all hoped, I suspect, that the ideal explanatory text would be a good deal simpler than Railton had conceived it to be, but it did not work out that way. It was clear to me long ago, for example, in elaborating the statistical relevance model, that an objectively homogeneous partition, based on all objectively relevant factors, would usually have a horrendous number of cells, and, consequently, that a complete S-R explanation would be frightfully complex. However, if it is only the ideal text that is so forbidding, the situation is not so hopeless. The ideal explanatory text contains all of the objective aspects of the explanation; it is not affected by pragmatic considerations. It contains all *relevant* considerations.

When we turn to explanatory information, pragmatic considerations immediately loom large. What part of the ideal explanatory text should we try to illuminate? Whatever is *salient* in the context under consideration. That depends upon the interests and the background knowledge of whoever seeks the explanation. It depends upon the explanation-seeking why-question that is posed. Pragmatic considerations must not be seen as total determinants of what constitutes an adequate explanation, for whatever explanation is offered must contain explanatory information that coincides with something or other in the ideal explanatory text. The ideal explanatory text determines what constitutes explanatory information, and distinguishes it from explanatory misinformation. *Relevance* is a matter of objective fact; *salience* is a matter of personal or social interest. Thus, I should be inclined to say, the putative astrological explanation of President John F. Kennedy's death—discussed in §4.4—can be ruled out because it fails to coincide with any part of the ideal explanatory text. The configuration of stars and planets and satellites at the time of Kennedy's birth is (I firmly believe) irrelevant to the date of his assassination.

Looking at the pragmatics of explanation in this way, we can account for the rejections and asymmetries of explanation, which van Fraassen takes to be crucial (see Railton 1981, 248; fn 15 explicitly refers to van Fraassen). We can evaluate the presuppositions of why-questions to see if they are objectively satisfied. We can take necessary steps to determine—on the basis of contextual factors—just exactly what why-question is being posed. To do so we must specify van Fraassen's contrast class. We can consider the background knowledge of the individual who poses the question to ascertain what is missing—what knowledge gaps need to be filled if that person is to achieve scientific understanding—what aspect of the ideal text needs to be exhibited. We would also take into account the capacity of the questioner to assimilate scientific information in order to determine the depth and detail appropriate for that individual.

Given Railton's conception of explanatory information and the ideal explanatory text, it seems to me that much of the longstanding battle—going back to the early skirmishes between the logical empiricists and the ordinary language philosophers—can be resolved. I see this achievement as one foundation upon which a new consensus might be erected.

Railton refers to his general conception of scientific explanation as the *nomothetic account*; he compares it to the received view in the following way:

> Where the orthodox covering-law account of explanation propounded by Hempel and others was right has been in claiming that explanatory practice in the sciences is in a central way *law-seeking* or *nomothetic*. Where it went wrong was in interpreting this fact as grounds for saying that any successful explanation must succeed either in virtue of explicitly invoking covering laws or by implicitly asserting the existence of such laws. It is difficult to dispute the claim that scientific explanatory practice—whether engaged in causal, probabilistic, reductive, or functional explanation—*aims* ultimately (though not exclusively) at uncovering laws. This aim is reflected in the account offered here in the structure of ideal explanatory texts: their backbone is a series of law-based deductions. But it is equally difficult to dispute the claim that many proffered explanations succeed in doing some genuine explaining *without* either using laws explicitly or (somehow) tacitly asserting their existence. This fact is reflected here in the analysis offered of explanatoriness, which is

treated as a matter of providing accurate information about the relevant ideal explanatory text, where this information may concern features of that text other than laws. (1981, 248–49)

In choosing to call his account *nomothetic* Railton emphasizes the role of laws. I think it equally deserves to be called *mechanistic*. We have already noted the crucial role of mechanisms in connection with the D-N-P model. In his 1981 article he remarks more generally on the standards for judging accounts of explanation:

The place to look for guidance is plainly scientific explanatory practice itself. If one inspects the best-developed explanations in physics or chemistry textbooks and monographs, one will observe that these accounts typically include not only derivations of lower-level laws and generalizations from higher-level theory and facts, but also attempts to *elucidate the mechanisms* at work. Thus an account of alpha-decay ordinarily does more than solve the wave-equation for given radionuclei and their alpha-particles; it also provides a model of the nucleus as a potential well, shows how alpha-decay is an example of the general phenomenon of potential-barrier penetration ("tunnelling"), discusses decay products and sequences, and so on. Some simplifying assumptions are invariably made, along with an expression of hope that as we learn more about the nucleus and the forces involved we will be able to give a more realistic physical model. It seems to me implausible to follow the old empiricist line and treat all these remarks on mechanisms, models, and so on as mere *marginalia*, incidental to the "real explanation", the law-based inference to the explanandum. I do not have anything very definite to say about what would count as "elucidating the mechanisms at work"—probabilistic or otherwise—but it seems clear enough that an account of scientific explanation seeking fidelity to scientific explanatory practice should recognize that part of scientific ideals of explanation and understanding is a description of the mechanisms at work, where this includes, but is not merely, an invocation of the relevant laws. Theories broadly conceived, complete with fundamental notions about how nature works—corpuscularianism, action-at-a-distance theory, ether theory, atomic theory, elementary particle theory, the hoped-for unified field theory, etc.— not laws alone, are the touchstone in explanation. (1981, 242)

In view of these comments, and many others like them, I am inclined to consider Railton's account primarily mechanistic and secondarily nomothetic.

Some time after I had seen his 1981 article, I happened to fall into conversation with Hempel—who was a colleague in Pittsburgh at the time—about Railton's work. He mentioned that he had a copy of Railton's dissertation (1980) and offered to lend it to me. After looking at parts of it rather quickly, I realized that I had found a treasure. Obtaining a copy of my own, I studied it at length, and made considerable use of it in conducting the Minnesota Institute on Scientific Explanation in 1985. In a general survey article such as this I cannot attempt a full summary of its contents, but a little more should be said to indicate the scope of the theory he develops.

In his dissertation Railton distinguishes three kinds of explanations of particular facts. In response to various difficulties with D-N explanations of particular facts as treated by the received view, Railton maintains that an appeal to causality is required in many cases. He characterizes the *ideal causal nomothetic explanatory text for particular facts* and concludes that it "includes an account of the causal mechanisms involved at all levels, and of all the strands in the causal network that terminates in the explanandum" (1980, 725). However, he resists the notion that all particular-fact explanations are causal. There are, as we have seen, probabilistic explanations of particular facts, and we have discussed the appropriate type of ideal explanatory text. As an advocate of the propensity interpretation of probability, he regards probabilistic explanations as dispositional. But he is willing to take the concept of disposition in a sense broad enough to include dispositions of strengths zero and unity; consequently, he introduces, in connection with his second type of particular-fact explanation, the notion of the *ideal explanatory text for dispositional particular-fact explanations*, of which the D-N-P model provides a special case (1980, 735–36). The third type of particular-fact explanation, for Railton, is structural. Such explanations appeal typically to such laws of coexistence as the Pauli exclusion principle, Archimedes' principle, laws of conservation, and so on. Maintaining that such explanations cannot plausibly be construed as disguised causal explanations, he also introduces the notion of the *ideal explanatory text for structural particular-fact explanations* (1980, 736–39). Explanations of this type – like causal and dispositional explanations – involve essential reference to the underlying mechanisms.

Recognizing that actual explanations of particular occurrences may involve elements of more than one of the foregoing, Railton also gives us the notion of a *nomothetic ideal encyclopedic text* (1980, 739). "Harmonious co-operation among the elements of an encyclopedic text is possible because all three forms reflect a conception of explanation that we may intuitively describe as this: an explanation should invoke the factors, laws, etc., that actually *bring about* or *are responsible for* the explanandum (once these relations are broadened beyond the purely causal), i.e., the explanations should show what features of the world the explanandum is *due to*" (1980, 739–40).

Because of certain differences in terminology among the three of us, perhaps it would be helpful to say something about the relationships among Coffa, Railton, and me. First, it should be noted, I am an advocate of probabilistic causality, so when I speak of causal explanation it is explicitly intended to include probabilistic explanation. Both Coffa and Railton use the term "cause" in the narrower deterministic sense. Furthermore, Coffa and Railton advocate the propensity interpretation of probability, while I reject the notion that propensities satisfy the axioms of probability, and hence that there is any such thing as a propensity interpretation. However, I do believe that the concept of a probabilistic propensity is an extremely useful notion; I would identify it with probabilistic cause. Hence, not to put too fine a point on it, Coffa, Railton, and I all agree on the general idea of a causal or dispositional type of explanation that coincides with Railton's causal and dispositional types. Up to this point, the differences are mainly terminological. A major difference arises, however, between Coffa on the one hand and Railton and me on the other, for both of us place far more importance upon an appeal to mechanisms than does Coffa. On another point Railton differs from Coffa and me; neither of us accords a distinctive place to Railton's structural explanations. My *constitutive explanations* bear some resemblance, but I regard them as fundamentally causal (1984, 270–76), whereas Railton explicitly denies that his structural explanations are. While I am still inclined to think that many of Railton's structural explanations can be analyzed causally, I am far from confident that all can. That point seems particularly clear with regard to such quantum mechanical examples as the Pauli exclusion principle, but, as we know, the nature of quantum mechanical explanation is deeply perplexing (W. Salmon 1984, 242–59). Nevertheless, had I read Railton's dissertation before the publication of my book, I would have devoted considerably more attention to structural explanation.

Railton's overall account is not confined to particular-fact explanations; he also offers an account of theoretical explanation. In summarizing his approach he remarks,

> It was argued that by beginning with particular-fact explanation, we had not prejudiced the account against theoretical explanation, for, in fact, theoretical explanation had been involved all along in the full development of ideal explanatory texts. As might have been expected, the nomothetic account recognizes that regularities and laws may have causal, dispositional, or structural elements in their explanations, and these elements have virtually the same ideal forms as in particular-fact explanation. Thus the same requirements of true ingredients, basic covering-law structure, an account of mechanisms, thorough-going theoretical derivation, asymmetries, etc., apply in theoretical explanation. The nomothetic account thereby preserves the estimable unity of theoretical and particular-fact explanation that is characteristic of covering-law approaches to explanation. (1981, 746–47)[28]

Railton's dissertation offers a deep and sensitive treatment of a wide range of problems, issues, and views regarding the nature of scientific explanation. Although it does not attempt to resolve some of the most fundamental problems we have encountered in our discussions of explanations — such problems as the nature of laws, the analysis of causality, the nature of mechanisms, the notion of a purely qualitative predicate—it is a rich source of philosophical insight on the nature of scientific explanation. While I do not by any means agree with all of his views, I do believe that anyone who is seriously interested in philosophical work on scientific explanation should study his dissertation with care.

4.7 Aleatory Explanation: Statistical vs. Causal Relevance

In the early part of the third decade—when I was busily expounding the statistical-relevance model—I was aware that explanation involves causality, but I hoped that the required causal relations could be fully explicated by means of such statistical concepts as screening off and the conjunctive fork. A decade later, I was quite thoroughly convinced that this hope could not be fulfilled (W. Salmon 1980; see also 1984, chap. 7). Along with this realization came the recognition that statistical relevance relations, in and of themselves, have no explanatory force. They have significance for scientific explanation only insofar as they provide evidence for causal relations. By 1984 (34–47) they had been relegated to *the S-R basis* upon which causal explanations can be founded. Causal explanation, I argued, must appeal to such mechanisms as causal propagation and causal interactions, which are not explicated in statistical terms.

The question arises of what to do with the S-R basis. Given the fact that it will often be dreadfully complex, we might consign it to Railton's ideal explanatory text, recognizing that it is the sort of thing that will not often be spelled out explicitly. We will refer to parts of it when we need to substantiate causal claims.

Another sensible approach to this problem is, in the words of Frank Lloyd Wright, "Abandon it!"[29] This is the tack taken by Paul Humphreys in articulating his theory of *aleatory explanation* (1981, 1983). A basic difference between Humphreys's model and other models of probabilistic or statistical explanation extant at the time is that all of the latter require one or more probability values to appear explicitly in the completed explanation. Although knowledge of probabilities is used in constructing aleatory explanations, values of probabilities are absent from the explanation itself.

According to Humphreys, factors that are causally relevant are also statistically relevant. Just as there are two kinds of statistical relevance—positive and negative—so also are there two kinds of causes. Causes that tend to bring about a given effect are *contributing causes*; those that tend to prevent the effect are *counteracting causes*. The canonical form for an aleatory explanation is " 'A because Φ, despite Ψ', where Φ is a non-empty set of contributing causes, Ψ is a set, possibly empty, of counteracting causes, and A is a sentence describing what is to be explained." It is assumed that Φ and Ψ include all known causally relevant factors. The set of contributing causes must not be empty, for we have no explanation at all if only counteracting causes are present.

Consider a modified version of one of Humphreys's examples. Suppose that a car has gone off a concrete road at a curve and that the only known conditions that are causally relevant are the fact that the driver was fully alert and the car was traveling at an excessive speed. The first is obviously a counteracting cause of the accident; the latter a contributing cause. We might then say that the car went off the road because it was traveling too fast, despite the fact that the driver was alert. There are, of course, many other statistically relevant factors, and if they are taken into account the probability of the explanandum will change. But as long as there are no other factors that screen off excessive speed, or render it a counteracting cause, its status as a contributing cause is unchanged. Similarly, if there are no other factors that screen off driver alertness or transform it into a contributing cause, its status as a counteracting cause holds.

Suppose now we find out that, in addition, visibility was clear, but there was sand on the road at the curve. The first of these is a counteracting cause of the accident, and the second is a contributing cause. We may add both to the original explanation, transforming it into the following: The car went off the road, despite the fact that the driver was alert and visibility was clear, because the car was traveling too fast and there was sand on the road at the curve. The result is that the first explanation, though incomplete, is correct; we have not been forced to retract any part of the original. If, in contrast, we had been trying to construct an S-R explanation of the same phenomenon, we would have been required to retract the original statistical relevance relations and replace them with others.

In constructing aleatory explanations, we must be aware of the possibility that the introduction of an additional causally relevant factor may change a contributing cause into a counteracting cause or vice-versa, or it may render a cause of either type irrelevant. Suppose we learn that there was ice on the road at this curve. This would change the presence of sand on the road from a contributing cause to a counteracting cause, for if the road were icy the sand would tend to prevent the car from skidding. In this case, the additional factor would force us to reject the former explanation, even as a correct partial explanation.

When constructing aleatory explanations, we are aware of the danger of a defeating condition—i.e., a condition, such as ice on the road in the preceding example, that transforms a contributing cause into a counteracting cause or vice-versa. By careful investigation we try to assure ourselves that none are present. Establishing such a result would be analogous to verifying one of Coffa's extremal clauses (§3.3). Just as we might check for the absence of compressing forces before applying the law of thermal expansion to a heated iron bar, so also would we check for road surface conditions that could transform sand on the road from a contributing to a counteracting cause before offering an explanation of the car leaving the road.

Humphreys's theory of aleatory explanation falls clearly within the ontic conception; it constitutes a valuable contribution to our understanding of causal explanation, where probabilistic causes are included. Inasmuch as its fullest articulation is contained in "Scientific Explanation: The Causes, Some of the Causes, and Nothing but the Causes," his contribution to this volume (Kitcher and Salmon 1989), I shall resist the temptation to discuss his theory at greater length and let him speak for himself.

4.8 Probabilistic Causality

In the third decade, as we saw, Coffa's theory placed the notion of a probabilistic disposition (what I would call a probabilistic cause) in a central position with respect to scientific explanation. During the same decade, a similar line of thought was pursued by Fetzer, whose theory was fully articulated early in the fourth decade. By the beginning of the fourth decade I was thinking seriously about how the S-R (statistical-relevance) model could be augmented by suitable causal considerations. At that time I was aware of only three theories of probabilistic causality that had been reasonably well worked out, namely, those of Reichenbach (1956), I. J. Good (1961–62), and Patrick Suppes (1970). Neither Reichenbach nor Suppes drew any connection between his theory of probabilistic causality and scientific explanation; Good used his mathematical definition of degree of explicativity (1977) in a key role in his probabilistic causal calculus, but without much philosophical elaboration. I surveyed these theories (1980) and pointed to severe difficulties in each; soon thereafter Richard Otte (1981) exhibited still greater problems in Suppes's theory. The primary moral I drew was that causal concepts cannot be fully explicated in terms of statistical relationships; in addition, I concluded, we need to appeal to causal processes and causal interactions. (For details see W. Salmon 1984, chaps. 5–7.)

Early in the fourth decade Fetzer and Nute (1979) published a new theory of probabilistic causality that was intended to play a crucial role in the explication of scientific explanation.[30] Fetzer's theory of scientific explanation is based on two cardinal principles: first, the interpretation of probability as a single-case propensity, and second, the inadequacy of extensional logic for the explication of such fundamental concepts as lawlikeness and causality. Single-case propensities are understood as probabilistic dispositions, indeed, as probabilistic causes. To elaborate a theory of scientific explanation that embodies these ideas, Fetzer and Nute construct a modal logic in which conditional statements embodying three kinds of special connectives are introduced, namely, subjunctive conditionals, causal conditionals (involving dispositions of universal strength), and probabilistic conditionals (involving dispositions having numerical degrees of strength). Those of the last type are statements of probabilistic causality.

The formal system they construct has 26 axiom schemas in all. The first six form the basis for the logic of subjunctive conditionals; the symbol for the connective is a fork ($\ni$). This logic is not different in kind from various well-known systems of modal logic. The next eight axiom schemas form the basis for the logic of universal causal conditionals; the symbol for this connective is the u-fork ($\ni_u$). There is nothing wildly nonstandard in this part of the system either. The final dozen axiom schemas form the basis for the logic of probabilistic causality. At this stage, however, instead of introducing one new connective symbol, the authors introduce a *nondenumerable infinity* of symbols ($\ni_n$), the n-forks, where n assumes the value of each real number in the unit interval. They then talk about establishing the well-formed formulas of this calculus "in the usual way."

What they have done is, on the contrary, most unusual. In the first place, the standard representation of real numbers is by means of sequences of digits, almost all of which are infinite sequences. It appears, then, that the vast majority of symbols of the form "$\ni_n$" are not single symbols, or finite strings of symbols, but infinitely long strings. Standard logical and mathematical languages, though they usually admit countable infinities of symbols, generally limit the well-formed formulas to finite length. In standard logical and mathematical languages there can exist names for no more than a denumerable subset of the real numbers. The use of a nondenumerable infinity of symbols and infinitely long well-formed formulas signal fundamental difficulties for the Fetzer-Nute proposed system.

In several of the axiom schemas, for example, the n-fork appears twice with the same subscript. If we look at a given formula containing two n-forks with their numerical subscripts—one that coincides with an axiom schema in all other respects than the equality or inequality of the two subscripts—we must determine whether the subscripts are identical to ascertain whether the formula is an axiom. This is done by comparing the two sequences digit by digit. If the two sequences are different, we will discover a discrepancy within a finite number of comparisons; but if they are the same we will never be able to establish that fact. There is no effective way of deciding, in general, whether two representations designate the same real number.

One of the basic virtues of a good axiomatic system is that it provides us with the capability of recognizing an axiom when we see one. If we have a finite number of finite formulas that capability is evident. In addition, given a finite number of *suitable* axiom schemas, each of which admits an infinite set of axioms, it is possible to provide an effective method for recognizing an axiom when we meet one. The Fetzer-Nute system, as presented, does not give us any such method. Unless the formation rules are spelled out in full detail, and a way is found to circumvent this problem, the probabilistic part of the calculus cannot get off the ground. Fetzer and Nute acknowledge that their nondenumerable infinity of well-formed formulas presents a difficulty when it comes to an attempt to establish the completeness of this part of their calculus (Fetzer and Nute 1979, 473; Fetzer 1981, 67), but I think the problem lies much deeper.

Another conspicuous feature of the probabilistic part of the calculus is that from twelve rather complicated axioms only four utterly trivial theorems are offered:

If p probabilistically implies q, then

    (a) p does not necessarily imply q, and p does not necessarily imply not-q;

    (b) p does not have a universal disposition to produce q, and p does not have a universal disposition to produce not-q;[31]

    (c) p is possible and not-q is possible;[32]

    (d) p and q are jointly possible.

Such a large tree should bear more fruit.

The foregoing criticisms are not meant to imply that all attempts to construct a probabilistic causal calculus within a nonextensional logic are bound to be futile or fruitless, but only that the Fetzer-Nute version needs further work if it is to succeed.

In the less formal portion of his work on scientific explanation, Fetzer offers searching discussions of Hempel's requirement of maximal specificity and of the homogeneity requirement I imposed on the S-R model. As we recall, in 1968 Hempel replaced his 1965 RMS with a revised version RMS*. Fetzer offers further revisions and offers RMS** which, he believes, deals adequately with the problems concerning relevance that I had raised (1981, 99). In addition, he offers a revised explication of homogeneity designed to escape certain difficulties he alleged to be present in mine.

Two main points of his critique are worth at least brief comment. The most serious, I think, involves "the mistaken identification of statistical relevance with explanatory relevance" (1981, 93), since "*statistically relevant* properties are not necessarily *causally relevant* (or *nomically relevant*) properties, and conversely . . . " (1981, 92). This is a criticism whose validity I have completely endorsed, as I have remarked repeatedly in foregoing sections. When the S-R model was first published I believed that causal relevance could be explicated entirely in terms of statistical relevance relations, making great use of the screening off relation. In that place, however, I did not attempt to carry out any such analysis. During the next several years I became increasingly skeptical about the viability of any such approach, arriving finally at the fairly strong conviction that it could not be done (1980). However, the details of the mechanisms of causality and their role in scientific explanation were not spelled out in full detail until (1984), three years after the publication of Fetzer's book.

Another one of Fetzer's basic arguments is that, according to my definition, the only objectively homogeneous reference classes are those that have only one member, and, consequently, the relative frequency of any attribute in such a class is, of logical necessity, either zero or one (1981, 86–94). Therefore, he claims, whereas I had accused Hempel's theory of an implicit commitment to determinism, in fact my account is at least equally guilty. As I had long been aware, the reference class problem is extremely serious for any theory of statistical explanation. In Hempel's account, RMS (later RMS*) was designed to deal with this problem; in mine, objective homogeneity was intended to handle it. In my initial presentations of the S-R model, early in the third decade, I handled the problem quite cavalierly, making passing reference to Richard von Mises's concept of a place selection. By the end of the third decade, I realized that it had to be taken very seriously. My first attempt at a detailed explication of objective homogeneity was badly flawed (1977b); an improved treatment, which I hope is more successful, was not published until 1984 (chap. 3). However that may be, I am still convinced that the concept of an *objectively homogeneous reference class* is legitimate and important. I maintain, for example, that the class of carbon-14 atoms is objectively homogeneous with respect to the attribute of spontaneous radioactive decay within 5730 years. If our concept does not fit cases of that sort, our explication must be at fault.

Obviously I cannot accept Fetzer's view that problems with objective homogeneity are merely unfortunate consequences of adopting of the *extensional* frequency interpretation of probability; they are problems that can be solved and deserve to be solved. Moreover, as I have argued in detail (1979a), the single-case propensity interpretation does not escape what amounts to the same type of problem, namely, the specification of the chance set-up that is supposed to possess the probabilistic disposition, for on any propensity interpretation it is necessary to specify what counts as repeating the experiment. Fetzer addresses this problem by imposing *the requirement of strict maximal specificity*:

> An explanation of why an explanandum event . . . occurs is adequate *only if* every property described by the antecedent condition(s) is nomically relevant to the occurrence of its attribute property . . . . (1981, 125–26)

Nomic relevance is relevance by virtue of a causal or noncausal law. Nomic relevance is emphatically not to be identified with statistical relevance.

Fetzer uses this requirement in the formulation of his characterization of causal explanation:

> A set of sentences S, known as the "explanans," provides *an adequate nomically significant causal explanation of the occurrence of a singular event described by another sentence E, known as its explanandum, relative to [a given] language framework*, if and only if:
> (a) the explanandum is either a deductive or a probabilistic consequence of its explanans;
> (b) the explanans contains at least one lawlike sentence (of universal or statistical) 'causal' form that is actually required for the deduction or probabilistic derivation of the explanandum from its explanans;
> (c) the explanans satisfies the requirement of strict maximal specificity (RSMS) with respect to its lawlike premise(s); and

(d) the sentences constituting the explanation—both the explanans and the explanandum—are true, relative to the [given] language framework. (1981,126–27)

As a gloss on this formulation, Fetzer remarks, "If the law(s) invoked in the explanans are essentially universal, the logical properties of the relationship between the sentences constituting the explanans and its explanandum will be those of complete (deductive) entailment; while if they are essentially statistical, this relationship will be that of only partial (deductive) entailment. The logical relation, in either case, is *strictly deductive*" (1981, 127). In the following chapter, Fetzer offers an entirely parallel analysis of *nomically significant theoretical explanation*, a noncausal form of explanation.

Fetzer has offered a nonextensional explication which, by virtue of its appeal to partial entailment, bears striking resemblance to Mellor's nondeterministic version of the modal conception of scientific explanation. In commenting (above) on Mellor's views I expressed my strong doubts about the viability of the concept of partial entailment; these qualms apply equally to Fetzer's employment of it. Given the intensional analysis, however, Fetzer has a straightforward answer to the issue we raised in §4.3 concerning the relationship between descriptive and explanatory knowledge. He suggests that the distinction between *description and prediction*, on the one hand, and *explanation*, on the other, is that the former can proceed in an extensional language framework, while the latter demands an intensional language framework. It remains to be seen whether the intensional logic can be satisfactorily formulated.

Fetzer has offered an account of explanation that is, in an extended sense, deductive. His book is dedicated to Karl Popper. In the next section we shall consider a more orthodox Popperian approach to statistical explanation.

4.9 Deductivism

The thesis that all legitimate scientific explanations are deductive arguments has a long and proud history, going back at least to Aristotle. It has been reiterated by many philosophers, including John Stuart Mill in the nineteenth century, and Karl R. Popper in the early part of the twentieth century. During our four decades it has been advocated by Brodbeck, Stegmüller, and von Wright, as well as many others, including Popper and his followers. It is *not*, however, a view that Hempel ever explicitly held, for he steadfastly maintained that there are, in addition, explanations of the inductive-statistical type.[33]

Before the twentieth century, during the reign of classical physics and Laplacian determinism, deductivism with respect to scientific explanation was a natural and appealing view. One might have tolerated something like explanations of the inductive-statistical variety as long as it was clearly understood that they were incomplete, the appeal to probability or induction being merely a result of our ignorance of the full explanatory laws or facts. With the advent of quantum mechanics it became necessary to admit that determinism might well be false, and that deductive-nomological explanations of some important phenomena may be impossible in principle.

One possible response to these new developments in physics is simply to hold stubbornly to determinism; this reaction seems utterly anachronistic. New—as yet undiscovered—physical theories may eventually convince us that determinism is true; even so, we have no *a priori* guarantee that determinism will be reinstated. At this point in the history of physics our philosophical theories of explanation must leave open the possibility that the world is indeterministic.

Another possible response is to admit that quantum mechanics is an indeterministic theory, but then to deny that quantum mechanics furnishes explanations of any physical phenomena. This reaction also seems unwarranted. Quantum mechanics (including quantum electrodynamics and quantum chromodynamics) has had more explanatory success than any other theory in the history of science. Classical physics could not explain the distribution of energy in the spectrum of blackbody radiation; quantum mechanics provided the explanation. Classical physics could not explain the photoelectric effect; quantum mechanics could. Classical physics could not explain the stability of atoms; quantum mechanics could. Classical physics could not explain the discrete spectrum of hydrogen; quantum mechanics could. Unless one wants to retreat to the old position that science never explains anything, it seems implausible in the extreme to deny that quantum mechanics has enormous explanatory power.

Another avenue is, however, open to the deductivist. Without relinquishing the deductivist position, one can concede that quantum mechanics has great explanatory power, but that all of its explanations are of the type Hempel characterized as deductive-statistical. This means, of course, that quantum mechanics can furnish no explanations of particular facts; the only things it can explain are statistical laws. To take just one example, the deductivist who adopts this stance must maintain that, while quantum mechanics can explain how, in general, electrons are scattered by crystals, it cannot explain the particular patterns actually obtained by Davisson and Germer in their famous experiment.[34] I find it difficult to accept this conclusion. Given the fact that quantum mechanics governs the microstructure of all macroscopic phenomena, it would appear to lead to the conclusion that science cannot explain any particular occurrences whatever. As I argue in a paper that will appear in the first year of the fifth decade (1988), even if theoretical science could be shown to have no need of explanations of particular facts, it seems impossible to make the same claim for applied science. When, for example, questions of legal liability arise, we seek causal explanations (which may have probabilistic components) for such particular occurrences as the collapse of a tank holding vast quantities of oil or the contracting of lung cancer by a long-term heavy smoker of cigarettes. Philosophy of science that confines its attention to pure science, to the complete neglect of applied science, is, I suggest, severely biased.

Another maneuver available to the deductivist is to maintain that, although we cannot have a scientific explanation of a particular chance event, we can have an explanation of the fact that such an event has some particular probability. An easy way to see how it would work is to reconsider Railton's D-N-P model (discussed in section 4.6). In focusing on Railton's model, it is essential to keep in mind that he is not a deductivist. According to Railton, we can have explanations of chance events; such explanations consist of two parts: (1) a deductive argument, whose conclusion is that the event in question has a certain probability, and (2) a parenthetic addendum, which states that the event in fact occurred. If we were to accept the first part, while rejecting the second, we would be left with a deductive explanation of the fact that the explanandum-event has a particular probability.[35] Such a view leaves us in the position of having to say that science cannot explain what happens in the world; it can only explain why those things that do happen have certain probabilities. It can also explain why things that do not happen have certain probabilities of occurring.

When Richard Jeffrey challenged Hempel's claim that statistical explanations are arguments, he excepted certain "beautiful cases" in which the probability of occurrence is so great that there seems no point in giving any weight at all to their non-occurrence. One of his examples is the failure of a flat tire to reinflate spontaneously as a result of a jet of air formed by chance from the random motion of molecules in the surrounding air. Another example would be the melting of an ice-cube placed in a glass of tepid water. Although he allowed that such explanations may be construed as arguments, he did not, of course, claim that they are deductive. Railton maintained, on the contrary, that the mere fact that some probabilities are high and some not so high does not make a difference in principle between the "beautiful" and less attractive cases. If the latter are not arguments then neither are the former. My own view is that Railton is correct in this observation, but my opposition to the inferential conception of explanation is so deep that I may simply be prejudiced.

One staunch deductivist who has recently considered the problem of Jeffrey's "beautiful cases" is John Watkins, whose neo-Popperian *Science and Scepticism* appeared in 1984. Although he is a deductivist, he maintains that, while science cannot provide explanations of individual micro-events, it can furnish explanations of macro-events that consist of large aggregates of chance micro-events:

> There is a far-reaching analogy between the explanatory power of the deterministic theories of classical physics and the indeterministic theories of modern micro-physics . . . Both explain empirical regularities by appealing to higher level structural laws that are taken as immutable and absolute. In the

case of classical physics, such a law says that, given only that such-and-such conditions are satisfied, nothing whatever can prevent a certain outcome from following. In the case of microphysics, it says that, given only that such-and-such conditions are satisfied, nothing whatever can alter *the chance* that a certain outcome will follow. One could as well call the latter an "iron law" of chance as the former an "iron law" of nomic necessity. Both kinds of law, in conjunction with appropriate initial conditions, can explain empirical regularities and, indeed, singular macro-events (provided that the macro-event in question is the resultant of a huge aggregate of micro-events . . .). The analogy breaks down, however, when we come down to individual events at the micro-level. . . . I have argued that it is a mistake to call upon microphysics to *explain* an individual micro-event, such as the disintegration of a radon atom or the reflection of a photon; for if it really was a matter of chance which way it went, then the fact that it chanced to go this way rather than that simply defies explanation. What we *can* explain with the help of an appropriate indeterministic microphysical theory is why there was a precise objective probability that it would go this way. (1984, 246)

To support his claim that particular macro-events can be explained, Watkins discusses one of Hempel's familiar examples. Suppose we have a sample consisting of 10 milligrams of radon. We find, after 7.64 days (two half-lives of radon), that the sample contains 2.5 $\pm$ 0.1 milligrams of radon. Given the law of spontaneous radioactive decay and the fact that the original sample contained more than 10^{19} atoms, we can deduce the probability that at the end of 7.64 days 2.5 $\pm$ 0.1 milligrams of radon will remain. This probability is extremely close to unity; indeed, it differs from one by less than $10^{-(10^{15})}$.

> At this point, anyone, whether a deductivist, inductivist, or whatever, concerned with the nature of microphysics, faces a crucial question: what *physical* meaning should be given to vanishingly small values [of probability]? Should we interpret our disintegration law as allowing that the macro-outcome *might* fall outside the [given] interval? Should we interpret the laws of thermodynamics as allowing that patches of ice and wisps of steam *might* form spontaneously in an ordinary bathtub because of hugely improbable distributions of molecules? (1984, 243)

Following an approach to statistical laws that had been advocated by Popper (1959, 198–205), Watkins argues that in cases like Hempel's radon decay example, the statistical law is *physically equivalent* to a universal law, namely, given *any* 10 mg. sample of radon, it will contain 2.5 $\pm$ 0.1 mg. of radon after two half lives have transpired. If the statistical decay law is replaced by its universal surrogate, we can, of course, furnish a deductive explanation of the phenomenon. This is the foundation of Watkins's claim that, although we cannot explain individual micro-occurrences, we can provide explanations of macro-events that involve the behavior of extremely large numbers of micro-entities.[36]

Whether one accepts or rejects the Popper-Watkins thesis about deductive explanations of large aggregates of micro-events, the deductivist position — which rejects all such models of explanation as the I-S or S-R — is appealing because of its avoidance of the problems associated with maximal specificity, epistemic ambiguity, and explanations of improbable events. According to deductivism, given an indeterministic world, we must forego explanations of *what actually happens* (in many cases, at least); however, we can, by revealing the stochastic mechanisms, understand how the world works. I shall not go into greater detail about the merits of deductivism here, since Philip Kitcher's contribution to this volume (Kitcher & Salmon 1989) contains an extensive elaboration of that position.

There is, however, a residual problem with deductivism that merits attention. It has to do with the relationship of causality to explanation. In his celebrated work *The Cement of the Universe*, J. L. Mackie invites consideration of three machines that dispense candy bars (1974, 40–43). One of the machines is deterministic; the other two are indeterministic. Leaving aside some details that are inessential to our discussion, we can say that the deterministic machine gives you a candy bar if and only if you insert a shilling.

Among the indeterministic machines, the first gives a candy bar only if a shilling is inserted, but it sometimes fails to give one when the coin is put in. There is no deterministic explanation of these failures; they simply happen occasionally by chance. Suppose a shilling is inserted and the machine yields a candy bar. In this case, according to Mackie, putting the coin in the slot *causes* the candy bar to come out, for without the coin there would have been no candy bar. This is the *sine qua non* conception of causality; a cause is a necessary condition. I agree with Mackie about the causal relation here, and I would add that putting the coin in the slot *explains* the ejection of the candy bar.

The second indeterministic machine is the converse of the first. Whenever a shilling is inserted a candy bar is forthcoming, but occasionally, by chance, the machine ejects a candy bar when no coin is put in. The insertion of the coin is a sufficient, but not a necessary, condition of getting a candy bar. Suppose someone puts a shilling in this machine and receives a candy bar. According to Mackie it would be wrong to say that putting in the coin causes the candy bar to come out, for a candy bar might have been forthcoming even if no coin had been inserted. Again, I think that Mackie is right, and I would go on to say that the insertion of the coin does *not* furnish a (nonstatistical) explanation of the appearance of the candy bar.

If the intuition about these cases — shared by Mackie and me — is correct, it leaves the deductivist in an awkward position regarding scientific explanation. In the case of the first machine, we have identified the cause of the explanandum-event, but we cannot provide a D-N explanation, for the insertion of the coin is *not* a sufficient condition of getting a candy bar. In the case of the second machine, we can provide a D-N *explanation* of the explanandum-event, but there is no (nonprobabilistic) cause of it.

In my discussion of this point, to escape the blatant artificiality of Mackie's candy machines, I introduced two photon-detectors. The first never produces a click unless a photon impinges on it, but it fails to click in a small percentage of cases in which a photon impinges. The second detector never fails to click when a photon impinges, but occasionally it gives a spurious click when no photon is present. These detectors are strictly analogous to the candy-dispensing machines. With the first detector we have the cause of a click, but (according to the deductivist) no explanation. With the second detector we have (according to the deductivist) an explanation, but no cause. The deductivist, it seems to me, needs to come to terms with examples of this sort (W. Salmon 1988).

4.10 Explanations of Laws Again

There is one further point in Watkins's book that has important bearing on one of the recalcitrant problems we have encountered regarding deductive explanation. It will be recalled that Hempel and Oppenheim—in their notorious footnote 33—explained why they offered no account of deductive explanation of laws. Their difficulty was that Kepler's laws K could be deduced from the conjunction of those very laws with Boyle's law B, but this would surely fail to qualify as a bona fide explanation. The problem is to characterize precisely the distinction between deductions of that sort, which do not constitute explanations, and those derivations of regularities from more general laws that do constitute legitimate explanations.

According to Popperians, the fundamental aim of science is to produce and test bold explanatory theories. Watkins regards scientific theories as finite sets of axioms. The axioms must be logically compatible with each other and they must be mutually independent. He remarks, "It is rather remarkable that, although scientific theories are taken as the basic units by many philosophies and nearly all histories of science, there is no extant criterion, so far as I am aware, for distinguishing between a theory and an assemblage of propositions which, while it may have much testable content, remains a rag-bag collection" (1984, 204). As an answer to this problem he offers what he calls *the organic fertility requirement* (1984, 205). If a theory T contains more than one axiom, then it fulfills this requirement if it is impossible to partition the axiom set into two mutually exclusive and exhaustive nonempty subsets T' and T", such that the testable content of T is equal to the sum of the testable contents of T' and T". In other words, the axioms must work together to yield testable consequences that they cannot generate separately. With reasonable restrictions on what qualifies as testable content, it seems clear that the testable content of Kepler's laws and Boyle's law is no greater than the set theoretical union of the testable content of Kepler's laws and the testable content of Boyle's law. Kepler's three laws—K_1, K_2, K_3—presumably satisfy the organic fertility requirement, while the set consisting of those together with B would not.

The situation is somewhat complicated by the fact that any given theory can be axiomatized in many different ways. We could, for instance, form the conjunction of all four laws to give us a theory with just one axiom. To deal with this kind of move, Watkins provides five rules for "natural" axiom sets. Among them is one he designates as "Wajsberg's requirement"; it says, in part, "An axiom is impermissible if it contains a (proper) component that is a theorem of the axiom set" (1984, 208) This clearly disposes of the Hempel-Oppenheim example. Whether Watkins's requirements block all counterexamples is a question I shall not try to answer. If they do, then they can be used to solve the problem with which Michael Friedman was concerned. Theories that satisfy the organic fertility requirement and are axiomatized "naturally" serve to unify our scientific knowledge.

4.11 A Fundamental Principle Challenged

There is a principle that has long been considered a cornerstone in the theory of scientific explanation, namely, if a set of circumstances of type C on one occasion explains the occurrence of an event of type E, then circumstances of the same type C cannot, on another occasion, explain the nonoccurrence of an event of type E (or the occurrence of an event of a type E' that is incompatible with E). Since it has so often been taken as a first principle, let us call it "Principle I." If this principle were relinquished, it has been thought, the floodgates would be open to all sorts of pseudo-explanations that are scientifically unacceptable. Nevertheless, careful consideration of probabilistic or statistical explanation has led some authors to reject that principle.

The D-N model of scientific explanation clearly satisfies this principle, for it is impossible validly to deduce two incompatible propositions from any consistent set of premises. Hempel's I-S model also satisfies it, as long as the high-probability requirement is enforced; since the sum of the probabilities of two incompatible statements with respect to any given consistent body of evidence cannot exceed one, it is impossible for both to have high probabilities. Adherents of the modal conception are committed to Principle I, for to show that an occurrence is necessary obviously implies that any incompatible occurrence is impossible. In van Fraassen's pragmatic theory Principle I is embodied in the view that an explanation shows why the topic *rather than* any other member of the contrast class is true.

At various junctures in our discussion of statistical explanation, I have made reference to a symmetry principle, namely, if a given stochastic process gives rise to some outcomes that are highly probable and to others that are improbable, then we understand the improbable ones just as well (or as poorly) as the probable ones. If this symmetry principle is correct—and I believe it is—it puts us in a dilemma. It forces us to choose between abandoning Principle I or forgoing explanations of probabilistic outcomes. For if circumstances C explain the probable outcome E on many occasions, then the same circumstances C explain the improbable outcome E′ on some other occasions. For closely related reasons, both Stegmüller and von Wright rejected the claim that probabilistic explanations of particular occurrences are possible. Since I am not inclined to give up statistical explanations of single events, and since I am strongly opposed to any high probability requirement, I have (in company with Achinstein and Railton) rejected Principle I. Although there may be non-deductivists—such as Hempel in "Aspects of Scientific Explanation"—who reject the symmetry principle, thus embracing Principle I *and* statistical explanations of particulars, Principle I does seem to be one of the chief bludgeons of the deductivists (see, e.g., Watkins 1984, 246).

What are the hazards involved in the rejection of Principle I? We have often been warned—and rightly so—that science has no place for theological or metaphysical 'theories' that explain whatever happens. Suppose someone is critically ill. If the person dies, the loved ones explain it as owing to "God's will." If the person recovers, they explain it as owing to "God's will." Whatever happens is explained in terms of the will of the Almighty. However comforting such 'explanations' might be, they are vacuous because there is no independent way of determining just what God wills. To rule out 'explanations' of this sort we do not need to appeal to Principle I; it is sufficient to insist that scientific explanations invoke *scientific* laws and facts. Scientific assertions, including those employed for purposes of explanation, should be supported by evidence. If we are dealing with what Hempel and Oppenheim called "potential explanation," then the laws or theories involved must be capable of independent support.

At the outset of our story we mentioned the attitude of scientific philosophers toward Driesch's attempts to explain biological phenomena in terms of entelechies and vital forces. During the same era, Popper was severely critical of Freudian psychoanalytic theory and Marxian economics. In all of these cases the criticisms were directed against the empirical vacuousness of the theories involved. To the extent that the theories in question are, indeed, empirically vacuous, to that extent they *are* devoid of scientific explanatory import. Such theories differ radically from the basic statistical theories of contemporary physics. These statistical theories offer a range of possible outcomes from a given indeterministic situation, attaching a definite probability value to each. They are far from vacuous, and they are supported by a vast amount of empirical evidence. The fact that such theories are not deterministic does not rob them of explanatory power. Nowadays most philosophers would agree that they have the capacity to explain statistical regularities; the question is whether they can be used to explain individual occurrences. If Principle I—which is *not* needed to block vacuous explanations—is relinquished we can give an affirmative answer.[37]

Conclusion
Peaceful Coexistence?

We have arrived, finally, at the conclusion of the saga of four decades. It has been more the story of a personal odyssey than an unbiased history. Inasmuch as I was a graduate student in philosophy in 1948, my career as a philosopher spans the entire period. I do not recall when I first read the Hempel-Oppenheim essay, but I did discuss it in my classes in the early 1950s. My specific research on scientific explanation began in 1963, and I have been an active participant in the discussions and debates during the past quarter-century. Full objectivity can hardly be expected.

Certain large areas have been left virtually untouched. I have deliberately neglected such topics as explanation in biology, explanation in psychology, explanation in history, and explanation of human action largely because I am not well enough informed about the substantive subject matter. Certain related issues, especially causality, have been treated only in passing, because the literature is so vast, and because this essay is already far too long. I know that there are other important pieces of work, especially during the latter part of the four decades, that have not been mentioned. As the decades passed the volume of literature increased exponentially, and judgments about the recent past are probably more subjective than those pertaining to more remote times. My decisions about what to discuss and what to omit are, without a doubt, idiosyncratic, and I apologize to the authors of such works for my neglect. But I have made an honest effort to discuss the material that has mainly influenced the present state of the subject. That said, let me offer my personal appraisal of the situation.

5.1 Consensus or Rapprochement?

Is there a new consensus in philosophy of science regarding the nature of scientific explanation? Not to any very noticeable extent. There are, however, a few basic points on which there seems to be widespread agreement among those who are contributing actively to the philosophical discussion of the subject.

(1) At the beginning of the four decades, the view was rather widely held that it is no part of the business of science to provide explanations—that explanations can be found, if at all, only in the realms of theology and transcendental metaphysics. At present, virtually all philosophers of science of widely diverse persuasions agree that science can teach us, not only *that*, but also *why*. This is an important piece of progress. It is now generally acknowledged that one of the most important fruits of modern science is understanding of the world. We do not have to go outside of science to find it.

(2) There seems to be general agreement—but by no means as unanimous as opinion on the preceding point—that the 'received view' of the mid-1960s is not viable. Contemporary deductivism may retain some of the core of the 'received view', but it rejects the expansive attitude toward probabilistic explanations of particular facts, and recognizes the need for an improved view of the nature of deductive explanation.

(3) It is noteworthy that the Hempel-Oppenheim paper undertook to furnish a formal explication of at least one type of scientific explanation, providing the syntax and semantics of a formal language and offering precise logical definitions. Most subsequent treatments, including those of Hempel himself, have not aimed for that degree of formal rigor. There is, I think, a general tacit recognition that the kinds of tools employed by Hempel and Oppenheim are not especially fruitful for handling the problems encountered in this area. For my own part—but I emphasize that this is *not* part of any general consensus—I believe that what constitutes a satisfactory scientific explanation depends upon certain contingent facts about the universe, for example, what kinds of mechanisms actually operate in the physical world.

(4) There appears to be fairly wide agreement on the importance of the pragmatics of explanation, and on the recognition that this aspect was not accorded sufficient emphasis in the 'received view'.

Beyond these four points, I cannot think of any other areas in which consensus actually obtains. Nevertheless, another question should be raised, namely, is a new consensus *emerging* in philosophy of science? This question calls for a risky prediction, but I shall hazard a guess. It seems to me that there are at least three powerful schools of thought at present—the pragmatists, the deductivists, and the mechanists—and that they are not likely to reach substantial agreement in the near future.

Still another question should be raised. Even if there is no general consensus at present, and no bright prospects for one to emerge soon, is there a basis for a substantial degree of rapprochement among the differing viewpoints? Here I think an affirmative answer can be given.

Around the beginning of the second decade, when serious controversy about the nature of scientific explanation erupted, two points of view—those associated primarily with the names of Hempel and Scriven—appeared in opposition to each other. The Hempelian view, which became the received view, emphasized deductive subsumption, the covering law thesis, and nomic expectability. We have chronicled many of the vicissitudes it has suffered; the net result, I think, is that it emerges at the close of the fourth decade in the form of the unification thesis. This view, whose chief proponents are Friedman and Kitcher, holds that scientific understanding increases as we decrease the number of independent assumptions that are required to explain what goes on in the world. It seeks laws and principles of the utmost generality and depth. This is a view to which I believe Popperians also subscribe. As I have said above (§4.3), the explanatory goal of this approach is the construction of a coherent world picture, and the fitting of particular facts within this framework. On this conception, explanatory knowledge is not of some additional kind that transcends descriptive knowledge. Explanations serve to organize and systematize our knowledge in the most efficient and coherent possible fashion. Understanding, on this view, involves having a world-picture—a *scientific Weltanschauung*—and seeing how various aspects of the world and our experience of it fit into that picture. The world-picture need not be a deterministic one; nothing in this view precludes basic laws that are irreducibly statistical.

The unification approach, as defended recently by Watkins and pursued in this book (Kitcher & Salmon 1989) by Kitcher, is, I believe, viable. In Kitcher's terminology, it is a "top-down" approach. It fits admirably with the intuitions that have guided the proponents of the received view, as well as those that inspire scientists to find unifying theories. This is the form in which the epistemic conception of scientific explanation can flourish today.

From the beginning, the most prominent critics of Hempel—especially Scriven—stressed two themes, causality and pragmatics. Often the two were closely related, for many of them took causality to be inescapably context-dependent. As things developed, however, the emphasis upon causality and upon objective statistical relevance relations issued in the ontic conception of scientific explanation, in which the objective relations among events could be considered quite apart from pragmatic considerations. As this approach had developed by the close of the fourth decade, it became the causal/mechanical view that is advocated by—among others—Humphreys, Railton, and me.

Although it would be unfair for me to assume that Humphreys and Railton would agree, as I see it, this version of the ontic conception has developed into a view that makes explanatory knowledge into knowledge of the hidden mechanisms by which nature works.[1] It goes beyond phenomenal descriptive knowledge into knowledge of things that are not open to immediate inspection. Explanatory knowledge opens up the black boxes of nature to reveal their inner workings. It exhibits the ways in which the things we want to explain come about. This way of understanding the world differs fundamentally from that achieved by way of the unification approach. Whereas the unification approach is "top-down," the causal/mechanical is "bottom-up."

When we pause to consider and compare these two ways of looking at scientific explanation, an astonishing point emerges. These two ways of regarding explanation are *not incompatible* with one another; each one offers a reasonable way of construing explanations. Indeed, they may be taken as representing two different, but compatible, aspects of scientific explanation.[2] Scientific understanding is, after all, a complex matter; there is every reason to suppose that it has various different facets.

Let me illustrate this point by recounting an actual incident.[3] Several years ago, a friend and colleague—whom I shall call *the friendly physicist*—was sitting on a jet airplane awaiting takeoff. Directly across the aisle was a young boy holding a helium-filled balloon by a string. In an effort to pique the child's curiosity, the friendly physicist asked him what he thought the balloon would do when the plane accelerated for takeoff. After a moment's thought the boy said that it would move toward the back of the plane. The friendly physicist replied that *he* thought it would move toward the front of the cabin. Several adults in the vicinity became interested in the conversation, and they insisted that the friendly physicist was wrong. A flight attendant offered to wager a miniature bottle of Scotch that he was mistaken—a bet he was quite willing to accept. Soon thereafter the plane accelerated, the balloon moved forward, and the friendly physicist enjoyed a free drink.

Why did the balloon move toward the front of the cabin? Two explanations can be offered, both of which are correct. First, one can tell a story about the behavior of the molecules that made up the air in the cabin, explaining how the rear wall collided with nearby molecules when it began its forward motion, thus creating a pressure gradient from back to front of the cabin.[4] This pressure gradient imposed an unbalanced force on the back side of the balloon, causing it to move forward with respect to the walls of the cabin. Second, one can cite an extremely general physical principle, Einstein's *principle of equivalence*, according to which an acceleration is physically equivalent to a gravitational field. Since helium-filled balloons tend to rise in the atmosphere in the earth's gravitational field, they will move forward when the airplane accelerates, reacting just as they would if a gravitational field were suddenly placed behind the rear wall.

The first of these explanations is causal/mechanical. It appeals to unobservable entities, describing the causal processes and causal interactions involved in the explanandum phenomenon. When we are made aware of these explanatory facts we understand how the phenomenon came about. The second explanation illustrates the unification approach. By appealing to an extremely general physical principle, it shows how this odd little occurrence fits into the universal scheme of things. It does not refer to the detailed mechanisms. This explanation provides a different kind of understanding of the same fact. It is my present conviction that both of these explanations are legitimate and that each is illuminating in its own way.

If this assessment of the situation is correct, we have grounds for a substantial degree of rapprochement between two approaches to scientific explanation that have been in conflict for at least three decades. In so saying, I do *not* intend to suggest that there was no real opposition between these views from the beginning, and that all of the controversy has been beside the point. On the contrary, at the beginning of the second decade there was genuine disagreement between Hempel and Scriven, and there was subsequently genuine disagreement between Hempel and me on the relative merits of I-S and S-R explanations. Over the intervening years, however, both viewpoints have evolved and become more mature. This evolution has, I believe, removed the sources of earlier conflict and made possible an era of peaceful coexistence.

At this point there is a strong temptation to take a page from Carnap's *Logical Foundations of Probability* and announce that there are two concepts of scientific explanation—explanation$_1$ and explanation$_2$—both of which are perfectly legitimate, and which must not be confused with one another. The question is how to characterize them and their relations to each other. Let us identify explanation$_1$ with causal/mechanistic explanation. It could fairly be said, I believe, that mechanistic explanations tell us how the world works. These explanations are local in the sense that they show us how particular occurrences come about; they explain particular phenomena in terms of collections of particular causal processes and interactions—or, perhaps, in terms of noncausal mechanisms, if there are such things. This does not mean that general laws play no role in explanations of this kind, for the mechanisms involved operate in accordance with general laws of nature. Furthermore, it does not mean that explanations$_1$ can only be explanations of particular occurrences, for causal/mechanical explanations can be provided for general regularities. The causal/mechanical explanation offered in the case of the helium-filled balloon can be said to apply quite generally to lighter-than-air entities in similar circumstances. The molecular kinetic theory of gases provides a causal/mechanical explanation of such regularities as Boyle's law. Explanations$_1$ are *bottom-up* explanations, in Kitcher's terminology, because they appeal to the underlying microstructure of what they endeavor to explain.

Explanation$_2$ then becomes explanation by unification. Explanation in this sense is, as Friedman emphasized, global; it relates to the structure of the whole universe. Explanation$_2$ is *top-down* explanation. To reinforce the global character of explanation$_2$ we might restrict its applicability, as Friedman did, to explanations of regularities, but I do not think anything of much importance hinges on such a limitation.

If the foregoing suggestions are correct, we can reconcile the currently viable versions of the epistemic and ontic conceptions. That leaves the pragmatic view.

If the pragmatic approach is construed as the claim that scientific explanation can be explicated entirely in pragmatic terms, then I think our examination of Achinstein's, Bromberger's, and van Fraassen's work seriously undermines it. To give a correct characterization of scientific explanation we need to identify the kinds objective relevance relations that make an explanation *scientifically correct*. This task falls outside of pragmatics.

If, however, we see the pragmatic approach as illuminating *extremely important features* of explanation, without doing the whole job, we can fit it nicely into the foregoing account. As I remarked in §4.6, if we adopt Railton's concepts of *ideal explanatory text* and *explanatory information*, pragmatic considerations can be taken as determining what aspects of the ideal text are *salient* in a given context. The ideal text contains the objective *relevance relations* upon which correct explanations must be founded.

When we make reference to the ideal explanatory text, the question naturally arises whether it conforms to the unification conception or the causal/mechanical conception. The answer is, I think, both. If one looks at the main example Railton offered to illustrate his notion of a D-N-P ideal text—namely, alpha-decay of a uranium nucleus—we see that it contains elements of both. It appeals to the most general theory of quantum phenomena, but it also details the specific mechanism involved in the decay of one particular nucleus. Indeed, looking at that example, one is tempted to say that the top-down and bottom-up approaches are just two different ways of 'reading' the ideal explanatory text. Pragmatic considerations determine which way of 'reading' is appropriate in any given explanatory context. In the case of the friendly physicist, for example, an appeal to Einstein's equivalence principle would have been totally inappropriate; however, the causal/mechanical explanation might have been made intelligible to the boy and the other interested adults.

My remarks about the relationships among (what I see as) the three currently viable approaches to explanation have necessarily been brief and sketchy. They are reflections that have grown directly out of the writing of this essay. It remains to be seen whether they can be filled out in detail and applied more generally to a wide variety of examples of scientific explanation. If so, a new consensus can emerge from our present understanding of the topic.

5.2 Agenda for the Fifth Decade

Nothing could be more pleasing than the emergence of a new consensus along the lines just sketched, but whether that occurs or not, there are certain remaining problems that demand further serious consideration.

First among them, I believe, is a problem—or pair of problems—that has been with us from the beginning. This is the problem—raised, but not adequately answered, in the 1948 Hempel-Oppenheim paper—of lawlike statements and purely qualitative (or projectable) predicates. On this issue dissensus reigns. It is obviously crucial to anyone who adopts any sort of covering law view, but it is equally crucial to those who reject that conception if, like Scriven, they admit that laws have a role-justifying function.

Second, the problem of causality is still with us. It becomes especially critical for those who find an explanatory role for probabilistic causes. As my brief remarks have indicated, considerably more work is needed to clarify this concept. Recent interchanges with I. J. Good (W. Salmon 1988a), in addition to my critique of the Fetzer-Nute system, have convinced me that conflicting fundamental intuitions are rampant.

The foregoing two problems are hoary philosophical chestnuts on which much ink has been spilled. That situation will, no doubt, continue. The third problem is rather different, I believe. It has to do with quantum mechanical explanation. The chief source of the problem is the famous Einstein-Podolsky-Rosen paper.[5] It has, of course, been widely discussed, but not often explicitly in the context of scientific explanation. That paper, in effect, describes a thought-experiment and predicts its result. The issue has become more urgent recently as a result of Bell's theorem and the Aspect experiment on remote correlations (see, e.g., Mermin 1985 or Shimony 1988). Now there is an actual experimental outcome to explain. Opinions vary on the significance of these results. Some say that they are of little consequence, showing only that the results conform to the theoretical prediction of quantum mechanics. To the discredit of the received view, it would support that position and claim that, because of subsumption under a well-confirmed theory, the experimental result is explained. At the opposite end of the spectrum of opinion are some who say that Bell's theorem and Aspect's confirmation of the quantum mechanical prediction constitute the most important development in the history of physics. I cannot accept that assessment either. With N. David Mermin, I accept the *moderate view* that "anyone who isn't worried about this problem has rocks in their head."[6]

The situation, basically, is this. There are impressive remote correlations in the spin-states of photons that cannot be explained by local causal principles; action-at-a-distance appears to be manifested. The results can be derived from spin conservation, but it is nonlocal conservation, and we have no mechanism by which to explain it. I have no idea what an appropriate explanation would look like; we may need to know more about the microcosm before any explanation can be forthcoming. But I do have a profound sense that *something* that has not been explained needs to be explained. As I said at the close of *Scientific Explanation and the Causal Structure of the World*, "to provide a satisfactory treatment of microphysical explanation constitutes a premier challenge to contemporary philosophy of science" (279). That still strikes me as correct.

Notes

Introduction

1. See Rudolf Carnap (1966, reissued 1974, pp. 12–17) for interesting comments on the transition from the denial to the acceptance of the view that science can furnish explanations. This passage includes an interesting discussion of Hans Driesch's vitalism (including the appeal to entelechies), and of reactions to it by the logical positivists of the Vienna Circle.

2. It is possible, of course, to adopt a hypothetical or 'suspend the truth' attitude in which one asks how a particular event could be explained *if it were to occur*. This is *not* Velikovsky's attitude.

3. For a brief and accessible introduction see Gale (1981). The heading of this rather flamboyant article reads, "Certain conditions, such as temperature, were favorable to the emergence of life on earth. The anthropic principle argues the reverse: the presence of life may 'explain' the conditions." In the article Gale adds, "It is fair to say, however, that not all cosmologists and philosophers of science assent to the utility of the anthropic principle, or even to its legitimacy. Here I shall describe some of the ways in which the principle has been applied and let the reader judge its validity" (p. 154). For a more thorough and technical, as well as more recent, treatment see Barrow and Tipler (1986).

4. That feature, in itself, should give us pause, for it is an elementary logical fallacy to infer the truth of the premises from the truth of the conclusion of a valid deductive argument. As a matter of fact, neither Hempel nor I considers the traditional hypothetico-deductive schema an adequate characterization of scientific confirmation, but it seems to be so regarded by many people; see, for example, Braithwaite, *Scientific Explanation*, (1953 p. 9). Discussions of the shortcomings of the hypothetico-deductive method can be found in W. Salmon (1984, §30) or W. Salmon (1967, Chap. VII).

5. Samuel E. Gluck (1955) made a brief stab at the task, but it was insufficiently general and failed to take notice of such basic difficulties as the ambiguity of statistical explanation. One interesting feature of his article is the claim that, because of the inherently probabilistic character of physical measurement, even what we take to be D-N explanations in quantitative sciences are actually statistical explanations. This point has been a longstanding source of worry for deductivists.

6. There are, of course, precursors of the theory set forth in this article, including Aristotle and John Stuart Mill. An important twentieth-century precursor is Karl R. Popper, whose 1935 work sketches a version of D-N explanation, though not in the depth of detail of Hempel-Oppenheim. Moreover, Popper's book was not highly influential until the English translation (1959) was published.

The First Decade

1. For one explicit comment, see 1953, p. 347.

2. For a recent detailed and sophisticated discussion of the nature of laws see John Earman (1986, chap. 5).

3. For purposes of this discussion it is not necessary to draw a distinction between counterfactual and subjunctive conditionals. In the example, "If this table salt were placed in water it would dissolve," it does not matter whether or not the particular sample is at some future time placed in water.

4. I once made a very crude estimate of the amount of gold in the earth's seas, and it came out to be more than 1,000,000 kg; if sufficient resources were devoted to the project, somewhat more than 100,000 kg could be extracted from sea water and fashioned into a sphere.

5. It has been speculated that tachyons—particles that travel faster than light—exist, but on pain of contradiction with special relativity, they cannot be used to send messages.

6. In these remarks about the coextensiveness of laws and statements that support counterfactuals I am, of course, excluding counterfactuals based on logical truths and/or on definitions.

7. This may represent Braithwaite's fundamental view of the problem.

8. This point is closely related to the pragmatic view of John Stuart Mill, Charles Saunders Peirce, and David Lewis, according to which the laws are those generalizations that end up as fundamental principles in the ideal limit of scientific investigation. One crucial question with respect to this approach is whether ideal science must eventuate in a unique final form, or whether alternatives are possible. If alternatives are possible, this pragmatic resolution of the problem would not pick out a unique set of laws, for the future course of science is unpredictable. Consequently, the distinction between laws and nonlaws would be basically epistemic; the distinction would not be objective.

9. See the concept of resiliency in Brian Skyrms (1980).

10. A great deal of light will be shed on this issue by Bas van Fraassen's forthcoming book, *Laws and Symmetries*.

11. At this point in his discussion he is advancing this view for consideration, rather than asserting it, but this is precisely the conclusion he does draw at the end of this chapter, pp. 317–18.

12. Partly because of its somewhat opaque title and partly because of its formidable complexity this book received little attention. It was later reprinted under the more descriptive title, *Laws, Modalities, and Counterfactuals* (Berkeley & Los Angeles: University of California Press, 1976). In my foreword to this volume I attempted to survey the issues as they stood in 1976, and to provide a more accessible account of Reichenbach's major ideas.

13. Ernest Nagel, *The Structure of Science* (New York: Harcourt, Brace and World, 1961), Chap. 4.

14. This discussion was continued in (Goodman, 1947) and (Carnap, 1947a).

15. I am paraphrasing the Hempel–Openheim definitions, but I shall preserve their original numbering.

16. See W. Salmon, foreword, in Reichenbach (1947), pp. xxxii–xxxiii.

17. The expression "iff" is a standard abbreviation for "if and only if."

18. For the sake of a concrete interpretation of this counterexample, it will do no harm to restrict the range of our variables to humans.

19. In *no* sense is this remark intended as a historical comment; I do *not* mean to suggest that they were on the verge of sending it off for publication and spotted the problem just in the nick of time. I am merely drawing a parallel between this technical problem and that pointed out by Eberle, Kaplan, and Montague.

20. By the *conjunctive normal form* of a formula we mean an equivalent formula (containing just those sentential variables occurring essentially in the original) which is a conjunction each of whose terms is a disjunction of sentential variables or their negations.

21. It is a serious terminological error, I believe, to refer to the D-N model of scientific explanation as "*the* covering law model," for although it is one model that conforms to the general idea that explanation always involves subsumption under laws, it is by no means the only such model. Indeed, Hempel's I-S model is also a covering law model.

22. In conversation, when he visited the workshop at the Minnesota Center for Philosophy of Science.

23. It appears that Hempel published nothing on scientific explanation, beyond the Hempel-Openheim article, during this first decade of our chronicle.

24. For purposes of historical accuracy it should be noted that this chapter is based upon Braithwaite's 1946 presidential address to the Aristotelian Society. At about the same time Ernest Nagel (1956) produced an important study of functional explanation. This paper was prepared in 1953 as a reserach report at Columbia University. Neither Braithwaite nor Nagel makes any reference to the Hempel-Oppenheim paper in their discussions of teleological or functional explanation.

25. Like Braithwaite and Nagel, Scheffler also makes no mention of the Hempel-Oppenheim article. It might be remarked facetiously that if Scheffler had been sufficiently attuned to the historical situation he would have published this paper on teleology in 1957, and the 1957 critique of Hempel and Oppenheim in 1958.

26. Both Hempel and Nagel rely heavily on Merton's extensive discussion of *functional analysis*. Nagel's work (1956) contains a detailed analysis of Merton's treatment of this topic.

27. Although the first published version of the I-S model did not appear until three years later, it seems evident that Hempel had a pretty clear idea of what it would be like when he wrote the 1959 article.

28. Scriven (1959) would, I believe, be one example.

29. See, however, Scheffler (1957) for a discussion of the bearing of the referential opacity of intentional contexts in this connection.

30. Nagel (1961, chap. 1) lists four types of explanation: deductive, probabilistic, functional or teleological, and genetic, leaving open at that stage whether the latter two can be reduced to one or both of the former two.

31. John Canfield (1966) covers the discussion of functional explanation during a good deal of the first two decades of our story. It contains, in addition to classic papers by important authors, a clear and perceptive introduction by Canfield.

32. Wright (1973) contains important anticipations of the theory set out in his book.

The Second Decade

1. As we noted in §0.2, the most serious attempt before 1962 seems to be (Gluck, 1955), but it is much too sketchy to present a clearly articulated theory, and it fails to notice the serious difficulties involved in statistical explanation.

2. His book *The Philosophy of Science* (1953) contains many references to scientific explanation, but does not provide an explicit account of that concept. His *Foresight and Understanding* (1961) deos offer an explicit account.

3. In *Scientific Explanation* (1984, p. 130) I incorrectly attributed to Bromberger the view that all explanations can be appropriately requested by means of why-questions. I regret this error.

4. [Bromberger's footnote] 'Can think of no expression' and 'can think of no answer' as short for 'can imagine nothing, conjure up nothing, invent nothing, remember nothing, conceive nothing' . . . does not cover the familiar states of momentary amnesia during which one has the answer 'on the tip of one's tongue' but cannot utter it.

5. Broberger's footnote: "As an achievement term 'to explain' is also often used to credit people with certain scientific discoveries. 'Newton explained why the tides vary with the phases of the moon' may serve to mark the fact that Newton was the one who solved the riddle, who found the answer to a question with regard to which everybody had been in either a p-predicament or a b-predicament. To have explained something in this sense is to be one of the first to have explained it in the [previous] sense . . . to be one of the first to have been in a position to explain . . . it to a tutee."

6. See the charming dedication to Hempel under "NOTES" on p. 107.

7. These restrictions are contained in the definition of *general abnormic law* given in Bromberger (1966, p. 98).

8. For technical reasons an additional restriction — (4) the general rule completed by L has the property that if one of the conjuncts in the antecedent is dropped the new general rule cannot be completed by an abnormic law — is needed, but the main thrust if this account can be appreciated without it.

9. As Broberger notes explicitly, "unless" has to be contrued as the exclusive disjunction.

10. However, see Grünbaum (1963), Fetzer (1974), and Rescher (1963).

11. It should be recalled, from our discussion of functional or teleological explanation in §1.3, that Hempel did not base his qualms about explanations of these sorts on the problem of the temporal relation between the function and its goal.

12. I do not believe Bromberger ever published this precise example; his actual examples, which have the same import, are the height of a tower and the height of a utility pole to which a guy wire is attached.

13. Bas van Fraassen is an exception. In his 1980 work (pp. 132–34) he suggests that there are possible contexts in which such an explanation would be legitimate. We shall discuss his theory of explanation in §4.4 below.

14. This 'law' does, of course, make reference to particular entitites—earth, sun, and moon—but that, in itself, is not too damaging to the example. After all, in this respect it is just like Kepler's laws of planetary motion and Galileo's law of falling bodies. Like these, it qualifies as a derivative law, though not a fundamental law.

15. Scheffler (1957) subjected the symmetry thesis to searching criticism. To the best of my knowledge, this article is the first significant published critique of the Hempel-Oppenheim article.

16. See Grünbaum (1963) for a fuller discussion of this point.

17. As Philip Kitcher pointed out to me in a personal communication, Hempel could be defended against this example by arguing that "a *natural selection* explanation of the presence of a trait is really a deduction that the probability that the trait becomes fixed in a finite population is high."

18. This example is due to Henry Kyburg (1965). A variant of this example (due to Noretta Koertge) has unhexed table salt placed in holy water; and the 'explanation' of the dissolving is that the water was blessed and whenever table salt is placed in holy water it dissolves.

19. I offered one such method, based on Reichenbach's treatment of laws (Salmon, 1979).

20. In the introduction to this essay I noted that an earlier paper on statistical explanation by Samuel E. Gluck may qualify as a precursor.

21. I shall us the term "statistical law" to refer to factual generalizations, such as the chance of getting 6 if a standard die is tossed, or the probability that a carbon-24 nucleus will decay in 5730 years. One could say that statistical laws are empirical generalizations, provided it is clearly understood that this does *not* mean that only directly observable properties are involved. I shall use the term "law of probability" to refer to axioms and theorems of the mathematical calculus of probability. Such laws are not empirical and they do not have factual content.

22. I am not objecting to the use of idealizing simplifications in dealing with philosophical problems, but their limitations must be kept in mind.

23. Recalling the fact that, for the limiting frequency interpretation, the individuals have to be taken in some specific order, I must confess to a bit of queasiness here.

24. We shall find reasons below for doubting this assessment of the relative importance of D-S and I-S explanations.

25. Whether this statistical law asserts that r is the actual fraction of Fs within a finite class F, or the limiting frequency of Fs within an infinite class F, is an issue that need not concern us here.

26. If we are thinking of probabilities as limiting frequencies in infinite sequences, $P(G|F)$ may equal 1 and $P(G|F.H)$ may equal 0.

27. Hempel (1962) succumbed to this temptation, but he soon realized the essential difficulty, as he explains in (1965, p. 401, note 21).

28. I have slightly modified Hempel's notation, but not in any way that affects our discussion.

29. As we noted in discussing the precise Hempel-Oppenheim explication, the actual requirement for D-N explanation is that the explanans contain essentially at least one theory, where a theory may contain existential quantifiers and need not contain any universal quantifier. In the technical explication of a potential explanans, the explanatory theory must be true (since all theories, in the special sense of Hempel and Oppenheim, are, by definition, true). It would seem more reasonable to require only that the general premise be essentially generalized, for certainly we want to consider the explanatory power of theories (in the usual sense) that need not be true.

30. As we noted above, p. 16, although Braithwaite's *Scientific Explanation* contains a chapter devoted to explanations of laws, he does not come to grips with the fundamental problem noted by Hempel and Oppenheim.

31. Hempel published a revised version of RMS (1968), pp. 116–33, which he designated "RMS*." It embodies some technical revisions designed to overcome an objection by Richard Grandy, but these do not affect any of the philosophical issues we are discussing.

32. Linus Pauling's claims in this regard were receiving a good deal of publicity in the early 1960s (see Pauling, 1970).

33. I owe this example to the eminent geneticist Tracy Sonneborn. In a social conversation about hormones and sex, I asked him what would happen to a man who took oral contraceptives. Without a monent's hesitation he replied, "Well, he wouldn't get pregnant."

I am indebted to the eminent archaeologist William A. Longacre for furnishing further information regarding this question. He kindly sent me a copy of the 2 June 1987 issue of the tabloid *Sun* (vol. 5, no. 22) which carried the front page headline "WIFE FEEDS HUBBY BIRTH CONTROL PILLS TO STOP HIM CHEATING." According to the story (p. 35), "Determined to stop her husband from fooling around, a betrayed housewife came up with the idea of feeding him birth control pills—and the dumbfounded hubby was frightened out of his wits when the pills made him impotent." When she confessed that she had been concealing them in his food he became furious and "stormed out of the house and has since filed for divorce . . . on the grounds of mental and physical cruelty."

The Third Decade

1. It is worth noting that Hempel introduces the term "statistically relevant" in this article, but he does not use it to refer to a statistical relevance relation (as that concept is generally understood).

2. As the issue was addressed here, it was buried rather deep in a paper whose title gave hardly any hint that it dealt with this topic.

3. This paper is a much expanded and highly revised version of one presented at a University of Pittsburgh workshop in 1965.

4. I discuss these three conceptions—the epistemic, the modal, and the ontic—rather briefly in (1982, 1986); and also in greater detail in (1984, chaps. 1 and 4). We shall return to them in this essay when we get to the fourth decade.

5. The answer is obvious in the sense of Christopher Columbus. The story is told of a dinner attended by Columbus sometime after his first voyage to the new world. Some of the guests were belittling his accomplishment by suggesting that it was not all that difficult. Columbus requested an egg from the kitchen, which he passed around asking each guest to try to stand it on end. When all had failed and it came back to him, he gave it a sharp tap on the table, breaking in the shell at one end, whereupon it remained upright. "You see," he said, "it is easy after you have been shown how."

6. I have in mind chiefly Peter Railton and myself.

7. Suppose, for example, that we have a class of tosses consisting of tosses of many different coins. Suppose further that many of the coins are biased, some toward heads, others toward tails. Some of the coins are fair. Now if each of the biased coins is biased in the same degree toward heads or tails (as the case may be), if as many are biased toward heads as toward tails, and all of the coins are tossed equally often, then the probability of heads in the entire class will be ½. This class may be relevantly partitioned into three cells—tosses with coins biased toward heads, tosses with coins biased toward tails, and tosses of fair coins. The probability of heads in the third cell is equal to the probability of heads in the original class, but that does not mean that the partition is not relevant.

8. I am modifying the example somewhat for purposes of the present discussion.

9. These arguments were also contained in his doctoral dissertation, *Foundations of Inductive Explanation* (1973).

10. The treatment of the topic in this article was seriously flawed, and it is completely superseded by my 1984 work, (chap. 3).

11. We might roughly define violet light as that having wave-lengths in the range of approximately 3600–3900 Å and red light as that having wave-lengths in the range of approximately 6500–7100 Å. It is a *fact* that the normal human range of vision extends from about 3600 Å to about 7100 Å.

12. The addition to Hempel's RMS is italicized.

13. In developing the statistical-relevance model of scientific explanation, I employ a non-epistemically-relativized counterpart of RMS. It is the requirement that reference classes used in explanations of this sort be *objectively homogeneous*. A number of the considerations that enter into the foregoing revision of Hempel's RMS were developed in the attempt to provide an adequate characterization of objective homogeneity. See W. Salmon (1984, chap. 3).

14. This attempt was seriously flawed in several ways. It is completely superseded by my 1984 work (chap. 3).

15. The members of F must be taken in some order. If F is finite the order makes no difference; if F is infinite, choose some natural order, such as the temporal order in which the events occur or are discovered.

16. This requirement is analogous to Richard von Mises's principle of insensitivity to place selections in his definition of the *collective*.

17. Philip Kitcher criticizes this proposal as follows, "I suspect that the omnipresence of correlations will make for inhomogeneity everywhere. This [proposal] will not get around the spurious correlation problem. There's bound to be some (possibly hokey) property that all correlates of the Bs have." If he is right, there is more work to be done on the concept of objective homogeneity. One suggestion, essentially adopted by Coffa, is to restrict consideration to nomically relevant (not just statistically relevant) factors.

18. This 'law' needs more careful formulation. Innumerable automotive and plumbing problems have been occasioned by the fact that water expands when it is cooled.

19. I have treated these issues in some detail (1979).

20. This is a highly contentious issue. In a personal communication Philip Kitcher comments, "Here I think you are describing a different chance setup. It depends how far you go back in the causal chain. If the setup involves the producing machine, then the Laplacian description seems right. If we start with some already selected coin of unknown constitution, then it seems that [the other] is right. We don't have an ambiguous description *of the same situation.*" My answer to this ploy is that we have two different descriptions of two different *types* of chance setups. The event in question actually belongs to both. If we want to characterize *this chance setup*, we have to decide which type to assign it to.

In another personal communication Paul Humphreys writes, "Your argument here applies only to long-run propensity interpretations, and not to single case propensities. Under the single case account, the continued operation of the machine is irrelevant to each specific trial, and under either experiment you describe, the propensity will be either less than or greater than ½, although in the first experiment it will remain fixed, whereas in the second experiment it will change depending on which kind of disk is picked on each trial." My answer to Humphreys is that he has done what any good limiting frequency or long-run propensity theorist would do. He has picked the appropriate type of chance setup to characterize *this trial*.

21. Because Coffa did not develop the idea or spell out any details, I had completely missed the point and forgotten all about it until I very recently reread his disseration. Every important philosophical discovery has to have a precursor; Coffa plays that role for Humphreys.

22. Hempel's response to Rescher can be found in (Hempel, 1965, pp. 403–6).

23. This book contains a comprehensive bibliography of works on scientific explanation up to its date of publication.

24. Indeed, any but the simplest putative Hempelian explanation of an individual event, because it involves subsumption under one or more laws, can be construed as an explanation of a regularity. The explanation of the bursting radiator, for example, is an explanation of why *any* radiator of a similar type, subjected to similar conditions, will burst.

25. Kicher responds, of course, that the Hempel-Oppenheim criteria are too weak. Where they required merely that the primitive predicates be purely qualitative, Kicher seems to want to require that they signifiy natural kinds.

26. Kitcher will argue, of course that typically not all of the Hs will represent predicates that are projectable from their instances.

27. Kitcher responds, "I'm not sure you could project either generalization without having a background theory about organisms that would give you both at once—or that would tell you that one is true if and only if the other is."

28. In a personal communication Kitcher also rejects this example: "I'm not sure we could accept any of these without accepting them all. The reason is that the projection from a finite sample seems to depend on believing that the relevant predicates pick out crucial classes in nature. There's no basis for thinking of the large-small cases, for example, as a privileged class for projection without thinking that the class of *all* two-body systems is projectable. Sometimes, I believe, you can't make a *restricted* projection. You either go all the way or nowhere."

29. Philip Kitcher, "Explanation, Conjunction, and Unification," *Journal of Philosophy* LXXIII (1976), p. 209.

30. In a personal communication Kitcher reponds: "Not necessarily, if the conditions on acceptability concern the projectability of predicates." Kitcher's remarks about projectability in response to my criticism of Friedman are interesting and important. If, however, this is the avenue to salvation for Friedman's program, he clearly needs to add a theory of projectability to the proposal he has offered.

31. A.A. Few, "Thunder," *Scientific American* (July, 1975), pp. 80–90. [Kitcher's reference.]

32. The first two dogmas were elaborated in W. V. Quine's classic essay "Two Dogmas of Empiricism," (1951).

33. This is the position Jeffrey took (1969); however, I do not agree with his way of making the distinction between those that are and those that are not.

34. If these temporal relations are not obvious, think of a given photon approaching the vicinity of the flagpole. If it passes by the flagpole without being absorbed, it reaches the ground a little later than it passed by the flagpole. A companion photon, traveling alongside of the above-mentioned one in a parallel path that intersects with the flagpole, will be absorbed a little before the other reaches the ground. As a very rough rule of thmb, the speed of light is a billion ft/sec; the photons travel about one foot per nanosecond.

35. I offered one suggestion on how this could be done in "Postscript: Laws in Deductive-Nomological Explanation—An Application of the Theory of Nomological Statements" (1979).

36. Moves of this sort were made in Fetzer (1981).

37. Carnap argues in careful detail for his denial of rules of acceptance (1950, §44b, 50–51). We need not go into his reasons for this approach here, for we are concerned only with its consequences in this context.

38. It has been noted that the special theory of relativity, while prohibiting the acceleration of ordinary material particles from speeds below that of light to superluminal speeds, may not preclude the existence of particles—called *tachyons*—that always travel faster than light. It is generally agreed, I believe, that there is, at present, no empirical evidence for their existence. Moreover, the presumption is that, if they should exist, they would be incapable of serving as signals or transmitting information. See my 1975 work (2nd ed., 1980, pp. 105, 122–24) for a discussion of the problems that would arise if tachyons actually existed.

39. Many philosophers would analyze processes as continuous series of events. While this can surely be done, there is, in my opinion, no particular reason for doing so,and there are some significant disadvantages. See my 1984 work (pp. 140–41, 156–57) for a discussion of this issue. See also John Venn's nice remark quoted on p. 183.

40. A material object that is at rest in one frame of reference will, of course, be in motion in other frames of reference.

41. In speaking of interventions that produce marks, I do *not* mean to suggest that such occurrences must be a result of human agency. For example, light from a distant star may be marked as a result of passage through a cloud of interstellar gas that selectively absorbs light of certain frequencies while letting others pass through unaffected.

42. The pseudo-process consisting of the moving spot of light can be changed from white to red from some point onward by putting a red lens on the beacon, but that does not qualify as an intervention *in the pseudo-process* because it is done elsewhere.

43. I often put so-called scare-quotes around the word "process" when it is being used to refer to pseudo-processes, for many people might be inclined to withhold that term when they realize that the process in question is pseudo. Reichenbach called them "unreal sequences."

44. The Lone Ranger's horse Silver was described as "A fiery horse with the speed of light." For the image of the horse on the screen that is possible in principle.

45. In the infinitesimal calculus we do, to be sure, define the concept of *instantaneous velocity*, but that definition requires consideration of the position of the object at neighboring instants of time. See my 1970 work, which contains Russell's article on Zeno's Paradoxes, and my 1975 work (chap. 2) for a fuller account.

46. Because of my own disciplinary limitations I have not participated in this field.

47. Perhaps I should add, or a bit of teleological non-action, such as not removing a tree that just happened to grow where it did, because it provides some wanted shade.

48. I do not find Nagel's criticisms of Wright particularly weighty.

49. These include Andrew Woodfield (1976) and Michael Ruse (1973).

50. It should be recalled, as previously mentioned, that Hempel addressed the problem of functional explanation in such areas as anthropology, sociology, and psychology, as well as biology. Nagel confines his attention, in these lectures, to biology. It is *much* harder to argue against functional equivalents in these other fields than it is in biology.

The Fourth Decade

1. I have argued this issue briefly (1967, pp. 729–32), and in greater detail (1969).

2. Coffa had previously provided searching critiques of both of these models in his doctoral disseration (1973).

3. This theory was also presented in his doctoral dissertation; see §3.3 above.

4. Before his 1977 publication at any rate. As I remarked above, it is not clear how the I-S model can survive abandonment of the high-probability requirement.

5. I discussed these general conceptions at length (1984, chaps. 1 and 4), and more briefly (1982, 1985).

6. I have no qualms about considering "micrometer" a term of the observational vocabulary, for it is defined in terms of "meter," which in view of the International Prototype and zillions of meter sticks in homes, stores, laboratories, and shops throughout the world, is surely an observational term.

7. Perrin actually studied the Brownian movements of particles suspended in liquids, but that fact does not affect the final results as applied to gases.

8. In offering this response on behalf of the modal conception I am, of course, assuming that knowledge of physical necessity and impossibility are not fully contained within our descriptive knowledge of the world. I shall return to this issue shortly.

9. I take it that predictive knowledge qualifies directly as part of descriptive knowledge, for knowledge of what is going to happen in the future is surely included in descriptive knowledge of the world.

10. See J. L. Mackie (1974) for detailed discussion of this approach. Mackie, himself, does not adopt it.

11. See, for example, Carnap (1966, 1974) for an exposition of this view of causality in the context of scientific explanation.

12. See David Lewis, *Counterfactuals* (1973). As I mentioned above, Lewis himself attemps to break out of the circle by adopting the view that laws are the basic principles in an ideally complete science. In my foregoing remarks on this topic, it will be recalled, I raised the crucial question of the uniqueness of characterization—a problem Lewis acknowledges.

13. Some physicists have proposed the "many-worlds" interpretation of quantum mechanics to resolve the problem of measurement (see De Witt and Graham, 1973). I am not favorably incluined toward this interpretation for much the same reasons.

14. It should be recalled that the term "theory," as it occures in the Hempel-Oppenheim formal explication, does not involve any appeal to unobservables. It simply refers to general statements that may contain existential as well as universal quantifiers.

15. I shall dicuss the legitimacy of this claim to empirical knowledge of unobservables in §4.5.

16. To those who suggested that such feelings are observable, at least by the subject, they responded that introspection is not a scientifically acceptable kind of observation.

17. I have argued this point most explicitly in my 1985 publication; it is also discussed in my 1984 work (pp. 229–38).

18. See my *Scientific Explanation* (pp. 13–15) for a striking historical example.

19. There are, of course, unsatisfactory answers to questions, but I would suppose they should be called unsatisfactory explanations. Not all explanations are good explanations.

20. See Frolich (1980), especially the photographs on p. 154.

21. Indeed, the only detailed critiques of van Fraassen's treatment of explanation so far published of which I am aware are in Salmon (1984) and Kitcher and Salmon (1987). The criticisms expressed in my work are entirely different from those given in Kitcher and Salmon. Achinstein (1983) gives brief attention to van Fraassen's theory of explanation.

22. For van Fraassen's response to this argument, as well as to mine, see Churchland and Hooker (1985, pp. 297–300).

23. A rather good sample of the range of opinion can be found in (Churchland and Hooker, 1985) and in (Leplin, 1984). Essays by Fine, Laudan, Boyd, and McMullin can be found in the Leplin volume.

24. In his useful discussion, "Structural Explanation" (1978, pp. 139–47), Ernan McMullin stresses the causal foundations of structural explanations, but he too steers clear of quantum phenomena.

25. It is ideal in the sense not of something we should necessarily strive to realize, but, rather, of (a Platonic?) something that may not exist in the physical world. It is the sort of thing Laplace's demon might be able to realize, if it concerned itself with explanations.

26. Ironically, the published article contains a misprint, which I have corrected, in the first two words of this passage. Instead of "Is it" the original reads "It is." Has Freud struck yet again?

27. I suspect Railton might agree (see Railton, 1981, pp. 243–44).

28. At the close of §3 of his paper in this volume (Kitcher & Salmon 1989), Kitcher offers his critique of Railton and me on the issue of the role of mechanisms in explanations.

29. This was reportedly the terse recommendation of this famous architect when, many years ago, he was brought to Pittsburgh by the city fathers to give advice on what could be done to improve the city.

30. This material is incorporated, to a large degree, in Fetzer (1981). Nute was Fetzer's collaborator on the formal aspects of probabilistic causality, but not on the general theory of scientific explanation.

31. A probability of one is not equated with a universal disposition and a probability of zero is not equated with a universal negative disposition.

32. One would expect the stronger result, that p and not-q are jointly possible, to be derivable.

33. As we saw in §3.2, however, Hempel's doctrine of essential epistemic relativization of inductive-statistical explanation brought him dangerously close to the brink of deductivism.

34. I discuss this example and others in some detail (1984, pp. 111–20).

35. Railton remarks, incidentally, that he is not strongly opposed to discarding the parenthetic addendum (1981, p. 236). He remians, however, strongly opposed to the view that explanations are always arguments.

36. Popper (1959, pp. 198–205) offers a criterion for deciding when the number of micro-events is large enough to justify this kind of replacement; Watkins employs the same criterion in his account.

37. I have discussed the status of Principle I, and other closely related principles (1984, especially pp. 111–20).

Conclusion

1. It should not be supposed that all mechanical explanations appeal to unobservable mechanisms. One might explain the workings of some gadget solely in terms of observables to someone who had not noticed the mechanical relationships among them — e.g., the way in which squeezing the handbrake on a bicycle brings the bicycle to a stop. However, deeper scientific explanations do seem usually to invoke unobservables.

2. Gregory Cooper, who was an active participant in the Minnesota Workshop, independently recognized the compatability of these two approaches earlier than I did. His thought on this matter is contained in his doctoral disseration. I am happy to acknowledge his priority with respect to this point.

3. This little story was previously published (W. Salmon, 1981 pp. 115–25). I did not offer an explanation of the phenomenon in that article.

4. Objects denser than air do not move toward the front of the cabin because the pressure difference is insufficient to overcome their inertia.

5. Albert Einstein et al., "Can Quantum-Mechanical Description of Physical Reality Be Considered Complete?" *Physical Review* 47 (1935), pp. 777–80. In *The Book of Revelations* 13:18 it is said that the number of The Beast is 666; I believe it is 777.

6. N. David Mermin, "Is the Moon Really There When No One Is Looking? Quantum Theory and Reality," *Physics Today* (1985).

Chronological Bibliography

• Designates item on scientific explanation of special significance and/or receiving more than passing mention in this essay.

☐ Designates an item not directly on scientific explanation cited in this essay.

Standard textbooks on philosophy of science—many of which contain sections on scientific explanation, without contributing to research on the subject—are not included in this bibliography.

Antiquity

• Aristotle, *Posterior Analytics*.

Modern Prehistory

1843

• Mill, John Stuart, *A System of Logic*, first edition. London: John W. Parker.

1906

• Duhem, Pierre Maurice Marie, *La Théorie physique, son objet et sa structure*, first edition. Paris. [Second edition, 1914; English translation of second edition, *The Aim and Structure of Physical Theory*. Princeton: Princeton University Press, 1954.]

1913

☐ Perrin, Jean, *Les Atomes*. Paris: Alcan. [English translation New York: Van Nostrand, 1923.]

1919

Broad, C. D., "Mechanical Explanation and Its Alternatives," *Proceedings of the Aristotelian Society* 19, pp. 85–124.

1926

Ducasse, C. J., "Explanation, Mechanism, and Teleology," *Journal of Philosophy* 23, pp. 150–55.

1935

Broad, C. D., C. A. Mace, G. F. Stout, and A. C. Ewing, "Symposium: Mechanical and Teleological Causation," *Proceedings of the Aristotelian Society*, supplementary vol. 14, pp. 83–112.

Cornforth, K., "Symposium: Explanation in History," *Proceedings of the Aristotelian Society*, supplementary vol. 14, pp. 123–41.

☐ Einstein, Albert, B. Podolsky, and N. Rosen, "Can Quantum-Mechanical Description of Reality Be Considered Complete?" *Physical Review* 47, pp. 777–80.

• Popper, Karl R., *Logik der Forschung*. Vienna: Springer. [Imprint 1935, actually published in 1934]

1942

• Hempel, Carl G., "The Function of General Laws in History," *Journal of Philosophy* 39, pp. 35–48. [Reprinted in Hempel (1965); reprint pagination used for references in this book.]

1943

☐ Goodman, Nelson, "A Query on Confirmation," *Journal of Philosophy* 40, pp. 383–85.
• Rosenblueth, Arturo, Norbert Wiener, and Julian Bigelow, "Behavior, Purpose, and Teleology," *Philosophy of Science* 10, pp. 18–24.
White, Morton, "Historical Explanation," *Mind* 52, pp. 212–29.

1945

Feigl, Herbert, "Some Remarks on the Meaning of Scientific Explanation," *Psychological Review* 52, pp. 250–59. [Reprinted in Herbert Feigl and Wilfrid Sellars, eds., *Readings in Philosophical Analysis* (New York: Appleton-Century-Crofts).]

1946

Braithwaite, R. B., "Teleological Explanations: The Presidential Address," *Proceedings of the Aristotelian Society* 47, pp. i–xx.
Hospers, John, "On Explanation," *Journal of Philosophy* 43, pp. 337–46.
Miller, D. L., "The Meaning of Explanation," *Psychological Review* 53, pp. 241–46.

1947

☐ Carnap, Rudolf, "On the Application of Inductive Logic," *Philosophy and Phenomenological Research* 8, pp. 143–47.
☐ ——(a), "Reply to Nelson Goodman," *Philosophy and Phenomenological Research* 8, pp. 461–62.
Ginsberg, Morris, "The Character of Historical Explanation," *Proceedings of the Aristotelian Society*, supplementary vol. 21, pp. 69–77.
☐ Goodman, Nelson, "The Infirmities of Confirmation Theory," *Philosophy and Phenomenological Research* 8, pp. 149–51.
☐ ——(a), "The Problem of Counterfactual Conditionals," *Journal of Philosophy* 44, pp. 113–28. [Reprinted as chap. I of Goodman's *Fact, Fiction, and Forecast* (Cambridge, MA: Harvard University Press, first edition 1955).]
MacIver, A. M., "The Character of Historical Explanation," *Proceedings of the Aristotelian Society*, supplementary vol. 21, pp. 33–50.
Miller, D. L., "Explanation vs. Description," *Philosophical Review* 56, pp. 306–12.
☐ Reichenbach, Hans, *Elements of Symbolic Logic.* New York: Macmillan Co.
Strong, E. W., "Fact and Understanding in History," *Journal of Philosophy* 44, pp. 617–25.
Walsh, W. H., "The Character of Historical Explanation," *Proceedings of the Aristotelian Society*, supplementary vol. 21, pp. 51–68.

The First Decade

1948

• Hempel, Carl G., and Paul Oppenheim, "Studies in the Logic of Explanation," *Philosophy of Science* 15, pp. 135–75. [Reprinted in Hempel (1965); reprint pagination used for references in this book.]
☐ Quine, Willard van Orman, "Two Dogmas of Empiricism," *Philosophical Review* 60, pp. 20–43.

1949

Feigl, Herbert, "Some Remarks on the Meaning of Scientific Explanation," in Herbert Feigl and Wilfrid Sellars, eds., *Readings in Philosophical Analysis* (New York: Appleton-Century-Crofts), pp. 510–14.
Flew, Antony, "Psychoanalytic Explanation," *Analysis* 10, pp. 8–15.

Kneale, William, *Probability and Induction*. Oxford: Clarendon Press, pp. 92–110. [Reprinted, under the title "Induction, Explanation, and Transcendent Hypotheses," in Herbert Feigl and May Brodbeck, eds., *Readings in the Philosophy of Science* (New York: Appleton-Century-Crofts, 1953), pp. 353–67.]

☐ Merton, R. K., *Social Theory and Social Structure*. New York: Free Press. [Second edition, revised and enlarged, 1957.]

1950

☐ Carnap, Rudolf, *Logical Foundations of Probability*. Chicago: University of Chicago Press. [Second edition 1962.]

Cohen, Jonathan, "Teleological Explanation," *Proceedings of the Aristotelian Society* 51, pp. 225–92.

Nagel, Ernest, "Mechanistic Explanation and Organismic Biology," *Philosophy and Phenomenological Research* 11, pp. 327–38.

Taylor, Richard, "Comments on the Mechanistic Conception of Purposefulness," *Philosophy of Science* 17, pp. 310–17.

☐ Velikovsky, Immanuel, *Worlds in Collision*. Garden City, N. Y.: Doubleday & Co.

Wilkie, J. S., "Causation and Explanation in Theoretical Biology," *British Journal for the Philosophy of Science* 1, pp. 273–90.

1951

Deutsch, Karl W., "Mechanism, Teleology and Mind," *Philosophy and Phenomenological Research* 12, pp. 185–223.

Nilson, S. S., "Mechanics and Historical Laws," *Journal of Philosophy* 48, pp. 201–11.

1952

☐ Carnap, Rudolf, *The Continuum of Inductive Methods*. Chicago: University of Chicago Press.

• Gardiner, Patrick, *The Nature of Historical Explanation*. Oxford: Clarendon Press.

☐ Radcliffe-Brown, A. R., *Structure and Function in Primitive Society*. London: Cohen and West Ltd.

Strong, E. W., "Criteria of Explanation in History," *Journal of Philosophy* 49, pp. 57–67.

Watkins, J. W. N., "Ideal Types and Historical Explanation," *British Journal for the Philosophy of Science* 3, pp. 22–43.

1953

• Braithwaite, Richard Bevin, *Scientific Explanation*. Cambridge: Cambridge University Press.

Gregory, R. L., "On Physical Model Expanations in Psychology," *British Journal for the Philosophy of Science* 4, pp. 192–97.

Hofstadter, Albert, "Universality, Explanation, and Scientific Law," *Journal of Philosophy* 50, pp. 101–15.

Nagel, Ernest, "Teleological Explanation and Teleological Systems," in Sidney Ratner, ed., *Vision and Action* (Rutgers University Press: New Brunswick, NJ), pp. 192–223.

Skinner, B. F., "The Scheme of Behavior Explanations," in B. F. Skinner, *Science and Human Behavior* (New York: Free Press).

☐ Toulmin, Stephen, *The Philosophy of Science*. London: Hutchinson's University Library.

☐ Wittgenstein, Ludwig, *Philosophical Investigations*. New York: Macmillan.

1954

Brown, Robert, "Explanation by Laws in Social Science," *Philosophy of Science* 21, pp. 25–32.

Dray, William H., "Explanatory Narrative in History," *Philosophical Quarterly* 4, pp. 15–27.

☐ Malinowski, B., *Magic, Science and Religion, and Other Essays*. Garden City, N. Y.: Doubleday Anchor Books.

☐ Reichenbach, Hans, *Nomological Statements and Admissible Operations*. Amsterdam: North-Holland Publishing Co.

1955

Gallie, W. B., "Explanations in History and the Genetic Sciences," *Mind* 64, pp. 161–67. [Reprinted in Gardiner (1959), pp. 386–402.]

Gluck, Samuel E., "Do Statistical Laws Have Explanatory Efficacy?" *Philosophy of Science* 22, pp. 34–38.

☐ Goodman, Nelson, *Fact, Fiction, and Forecast*. Cambridge: Harvard University Press.

Hayek, F. A., "Degrees of Explanation," *British Journal for the Philosophy of Science* 6, pp. 209–25.

1956

Danto, Arthur C., "On Explanations in History," *Philosophy of Science* 23, pp. 15–30.

Ellis, Brian, "The Relation of Explanation to Description," *Mind* 65, pp. 498–506.

Feigl, Herbert, and Michael Scriven, eds., *Minnesota Studies in the Philosophy of Science*, vol. I. Minneapolis: University of Minnesota Press.

Hospers, John, "What is Explanation?" in Antony Flew, ed., *Essays in Conceptual Analysis* (London: Macmillan & Co., Ltd), pp. 94–119.

Hutten, E. H., "On Explanation in Psychology and in Physics," *British Journal for the Philosophy of Science* 7, pp. 73–85.

• Nagel, Ernest, "A Formalization of Functionalism (With Special Reference to its Application in the Social Sciences)," in Ernest Nagel, Logic Without Metaphysics (Glencoe, Ill.: The Free Press), pp. 247–83.

☐ Reichenbach, Hans, *The Direction of Time*. Berkeley & Los Angeles: University of California Press.

1957

Argyle, Michael, "Explanation of Social Behaviour," in Michael Argyle, *The Scientific Study of Social Behaviour* (London: Methuen), chap. 3.

• Donagan, Alan, "Explanation in History," *Mind* 66, pp. 145–64. [Reprinted in Gardiner (1959).]

• Dray, William, *Laws and Explanation in History*. London: Oxford University Press.

Frankel, Charles, "Explanation and Interpretation in History," *Philosophy of Science* 24, pp. 137–55. [Reprinted in Gardiner (1959), pp. 408–27.]

Goldstein, L. J., "The Logic of Explanation in Malinowskian Anthropology," *Philosophy of Science* 24, pp. 156–66.

Madden, Edward H., "The Nature of Psychological Explanation," *Methodos* 9, pp. 53–63.

☐ Merton, R. K., *Social Theory and Social Structure*, 2nd ed., revised and enlarged. New York: Free Press.

• Scheffler, Israel, "Explanation, Prediction, and Abstraction," *British Journal for the Philosophy of Science* 7, pp. 293–309.

Watkins, J. W. N., "Historical Explanation in the Social Sciences," *British Journal for the Philosophy of Science* 8, pp. 104–17.

The Second Decade

1958

Feigl, Herbert, Michael Scriven, and Grover Maxwell, *Minnesota Studies in the Philosophy of Science* II. Minneapolis: University of Minnesota Press.

Goudge, T. A., "Causal Explanations in Natural History," *British Journal for the Philosophy of Science* 9, pp. 194–202.

☐ Hanson, Norwood Russell, *Patterns of Discovery*. Cambridge: Cambridge University Press.

• Hempel, Carl G., "The Theoretician's Dilemma," in Feigl et al. (1958), pp. 37–98. [Reprinted in Hempel (1965); reprint pagination used for references in this book.]

Papandreou, Andreas G., "Explanation and Prediction in Economics," *Science* 129, pp. 1096–1100.
Passmore, John, "Law and Explanation in History," *Australian Journal of Politics and History* 4, pp. 269–76.
Rescher, Nicholas, "On Prediction and Explanation," *British Journal for the Philosophy of Science* 8, pp. 281–90.
  □ ——(a), "A Theory of Evidence," *Philosophy of Science* 25, pp. 83–94.
• Scheffler, Israel, "Thoughts on Teleology," *British Journal for the Philosophy of Science* 9, pp. 265–84.
• Scriven, Michael, "Definitions, Explanations, and Theories," in Feigl, Scriven, and Maxwell (1958), pp. 99–195.
Skarsgard, Lars, "Some Remarks on the Logic of Explanation," *Philosophy of Science* 25, pp. 199–207.
Yolton, J. W., "Philosophical and Scientific Explanation," *Journal of Philosophy* 55, pp. 133–43.

1959

Beattie, J. H. M., "Understanding and Explanation in Social Anthropology," *British Journal of Sociology* 10, pp. 45–60.
Dray, William H., " 'Explaining What' in History," in Gardiner (1959), pp. 403–8.
Ebersole, F. B., and M. M. Shrewsbury, "Origin Explanations and the Origin of Life," *British Journal for the Philosophy of Science* 10, pp. 103–19.
• Gardiner, Patrick, ed., *Theories of History*. New York: The Free Press.
Hanson, Norwood Russell, "On the Symmetry Between Explanation and Prediction," *Philosophical Review* 68, pp. 349–58.
Harris, E. E., "Teleology and Teleological Explanation," *Journal of Philosophy* 56, pp. 5–25.
• Hempel, Carl G., "The Logic of Functional Analysis," in Llewellyn Gross, ed., *Symposium on Sociological Theory* (New York: Harper & Row). [Reprinted in Hempel (1965); reprint pagination used for references in this book.]
Joynt, C. B., and Nicholas Rescher, "On Explanation in History," *Mind* 68, pp. 383–88.
Pitt, Jack, "Generalizations in Historical Explanation," *Journal of Philosophy* 56, pp. 578–86.
• Popper, Karl R., *The Logic of Scientific Discovery*. New York: Basic Books. [English translation of Popper (1935).]
• Scriven, Michael, "Explanation and Prediction in Evolutionary Theory," *Science* 30, pp. 477–82.
• ——(a), "Truisms as the Grounds for Historical Explanation," in Patrick Gardiner, ed., *Theories of History* (Glencoe: The Free Press), pp. 443–75.
Stannard, Jerry, "The Role of Categories in Historical Explanation," *Journal of Philosophy* 56, pp. 429–47.
Sutherland, N. S., "Motives as Explanations," *Mind* 68, pp. 145–59.
Yolton, J. W., "Explanation," *British Journal for the Philosophy of Science* 10, pp. 194–208.

1960

Dodwell, P. C., "Causes of Behaviour and Explanation in Psychology," *Mind* 69, pp. 1–13.
  □ Hempel, Carl G., "Inductive Inconsistencies," *Synthese* 12, pp. 439–69. [Reprinted in Hempel (1965); reprint pagination used for references in this book.]

1961

Barker, Stephen, "The Role of Simplicity in Explanation," in Herbert Feigl and Grover Maxwell, eds., *Current Issues in the Philosophy of Science* (New York: Holt, Rinehart, and Winston), pp. 265–73. [See also comments by Wesley C. Salmon, Paul Feyerabend, and Richard Rudner, and Barker's replies.]
Bromberger, Sylvain, "The Concept of Explanation," Ph. D. dissertation, Princeton University.
• Eberle, R., D. Kaplan, and R. Montague, "Hempel and Oppenheim on Explanation," *Philosophy of Science* 28, pp. 418–28.

Feigl, Herbert, and Grover Maxwell, eds., *Current Issues in the Philosophy of Science*. New York: Holt, Rinehart, and Winston.

Good, I. J., "A Causal Calculus (I-II)," *British Journal for the Philosophy of Science* 11, pp. 305-18; 12, pp. 43-51. See also "Corrigenda," 13, p. 88.

• Kaplan, David, "Explanation Revisited," *Philosophy of Science* 28, pp. 429-36.

Mandelbaum, Maurice, "Historical Explanation: The Problem of 'Covering Laws'," *History and Theory* 1, pp. 229-42.

• Nagel, Ernest, *The Structure of Science: Problems in the Logic of Scientific Explanation*. New York: Harcourt, Brace & World.

Rescher, Nicholas, and Carey B. Joynt, "The Problem of Uniqueness in History," *History and Theory* 1, pp. 150-62.

Scriven, Michael, "Discussion: Comments on Weingartner," *Philosophy of Science* 28, p. 306.

Sellars, Wilfrid, "The Language of Theories," in Herbert Feigl and Grover Maxwell, eds., *Current Issues in the Philosophy of Science* (New York: Holt, Rinehart, and Winston), pp. 57-77.

Stover, R. C., "Dray on Historical Explanation," *Mind* 70, pp. 540-43.

• Toulmin, Stephen, *Foresight and Understanding*. Bloomington: Indiana University Press.

Weingartner, Rudolph H., "Discussion: Explanations and Their Justifications," *Philosophy of Science* 28, pp. 300-5.

——(a), "The Quarrel about Historical Explanation," *Journal of Philosophy* 58, pp. 29-45.

1962

Alexander, Peter, "Rational Behaviour and Psychoanalytic Explanation," *Mind* 71, pp. 326-41.

Bartley, W. W. III, "Achilles, the Tortoise, and Explanation in Science and History," *British Journal for the Philosophy of Science* 13, pp. 15-33.

Brodbeck, May, "Explanation, Prediction, and 'Imperfect' Knowledge," in Feigl and Maxwell (1962), pp. 231-72.

• Bromberger, Sylvain, "An Approach to Explanation," in R. S. Butler, ed., *Analytical Philosophy—Second Series* (Oxford: Basil Blackwell), pp. 72-105. [American edition, New York: Barnes and Noble, 1965.]

Chomsky, Noam, "Explanatory Models in Linguistics," in Ernest Nagel, Patrick Suppes, and Alfred Tarski, eds., *Logic, Methodology and Philosophy of Science*, Proceedings of the 1960 International Congress (Stanford: Stanford University Press), pp. 528-50.

Feigl, Herbert, and Grover Maxwell, eds., *Minnesota Studies in the Philosophy of Science* III. Minneapolis: University of Minnesota Press.

Feyerabend, Paul, "Explanation, Reduction, and Empiricism," in Feigl and Maxwell (1962), pp. 231-72.

• Grünbaum, Adolf, "Temporally-Asymmetric Principles, Parity Between Explanation and Prediction, and Mechanism vs. Teleology," *Philosophy of Science* 29, pp. 146-70.

• Hempel, Carl G., "Deductive-Nomological vs. Statistical Explanation," in Feigl and Maxwell (1962), pp. 98-169.

——(a), "Explanation in Science and in History," in Robert G. Colodny, ed., *Frontiers of Science and Philosophy* (Pittsburgh: University of Pittsburgh Press), pp. 7-34.

Kim, Jaegwon, "Explanation, Prediction, and Retrodiction: Some Logical and Pragmatic Considerations," Ph. D. dissertation, Princeton University.

Margenau, Henry, "Is the Mathematical Explanation of Physical Data Unique?" in Ernest Nagel, Patrick Suppes, and Alfred Tarski, eds., *Logic, Methodology and Philosophy of Science*, Proceedings of the 1960 International Congress (Stanford: Stanford University Press), pp. 348-55.

Passmore, John, "Explanation in Everyday Life, in Science, and in History," *History and Theory* 2, pp. 105-23.

Rescher, Nicholas, "The Stochastic Revolution and the Nature of Scientific Explanation," *Synthese* 14, pp. 200-15.

• Scriven, Michael, "Explanations, Predictions, and Laws," in Feigl and Maxwell (1962), pp. 170–230.
—— (a), "Discussion: Comments on Professor Grünbaum's Remarks at the Wesleyan Meeting," *Philosophy of Science* 29, pp. 171–74.
Treisman, Michel, "Psychological Explanation: The 'Private Data' Hypothesis," *British Journal for the Philosophy of Science* 13, pp. 130–43.

1963

Alexander, Peter, *Sensationalism and Scientific Explanation*. New York: Humanities Press.
Baker, A. J., "Historical Explanation and Universal Propositions," *Australasian Journal of Philosophy* 41, pp. 317–35.
Baumrin, Bernard H., ed., *Philosophy of Science: The Delaware Seminar*, vols. 1–2. New York: John Wiley & Sons.
Brandt, Richard, and Jaegwon Kim, "Wants as Explanations of Actions," *Journal of Philosophy* 60, pp. 425–35.
• Bromberger, Sylvain, "A Theory about the Theory of Theory and about the Theory of Theories," in Baumrin (1963, vol. 2), pp. 79–105.
Brown, Robert, *Explanation in Social Science*. Chicago: Aldine Publishing Co.
Dray, William H., "Historical Explanation of Actions Reconsidered," in Hook (1963), pp. 105–35.
Fain, Haskell, "Some Problems of Causal Explanation," *Mind* 72, pp. 519–32.
Grünbaum, Adolf, "Temporally Asymmetric Principles, Parity Between Explanation and Prediction, and Mechanism versus Teleology," in Baumrin (1963, vol. 1), pp. 57–96. [Substantially the same as Grünbaum (1962).]
Hempel, Carl G., "Explanation and Prediction by Covering Laws," in Baumrin (1963, vol. 1), pp. 107–33.
Henson, R. B., "Mr. Hanson on the Symmetry of Explanation and Prediction," *Philosophy of Science* 30, pp. 60–61.
Hesse, Mary, "A New Look at Scientific Explanation," *Review of Metaphysics* 17, pp. 98–108.
Hook, Sidney, ed., *Philosophy and History*. New York: New York University Press.
• Kim, Jaegwon, "On the Logical Conditions of Deductive Explanation," *Philosophy of Science* 30, pp. 286–91.
Morgenbesser, Sidney, "The Explanatory-Predictive Approach to Science," in Baumrin (1963, vol. 1), pp. 41–55.
Rescher, Nicholas, "Discrete State Systems, Markov Chains, and Problems in the Theory of Scientific Explanation and Prediction," *Philosophy of Science* 30, pp. 325–45.
—— (a), "Fundamental Problems in the Study of Scientific Explanation," in Baumrin (1963, vol. 2), pp. 41–60.
☐ Salmon, Wesley C., "On Vindicating Induction," *Philosophy of Science* 30, pp. 252–61.
• Scheffler, Israel, *The Anatomy of Inquiry*. New York: Alfred A. Knopf.
• Scriven, Michael, "The Temporal Asymmetry Between Explanations and Predictions," in Baumrin (1963, vol. 1), pp. 97–105.
—— (a), "The Limits of Physical Explanation," in Baumrin (1963, vol. 2), pp. 107–35.
—— (b), "New Issues in the Logic of Explanation," in Sidney Hook, ed., *Philosophy and History* (New York: New York University Press), pp. 339–61.
—— (c), "Review of Ernest Nagel, *The Structure of Science*," *Review of Metaphysics* 17, pp. 403–24.
Sellars, Wilfrid, "Theoretical Explanation," in Baumrin (1963, vol. 2), pp. 61–78.

1964

Canfield, John, "Teleological Explanations in Biology," *British Journal for the Philosophy of Science* 14, pp. 285–95.
Donagan, Alan, "Historical Explanations: The Popper-Hempel Theory Reconsidered," *History and Theory* 4, pp. 3–26. [Reprinted in Dray (1966).]
Gallie, W. B., *Philosophy and the Historical Understanding*. London: Chatto & Windus.

Gustafson, Don F., "Explanation in Psychology," *Mind* 73, pp. 280–81.

Jarvie, Ian C., "Explanation in Social Science," *Philosophy of Science* 15, pp. 62–72. [Review of Robert Brown, *Explanation in Social Sciences*.]

Kim, Jaegwon, "Inference, Explanation, and Prediction," *Journal of Philosophy* 61, pp. 360–68.

Martin, Michael, "The Explanatory Value of the Unconscious," *Philosophy of Science* 31, pp. 122–32.

Sorabji, Richard, "Function," *Philosophical Quarterly* 14, pp. 289–302.

• Taylor, Charles, *The Explanation of Behaviour*. London: Routledge and Kegan Paul.

Workman, Rollin W., "What Makes an Explanation," *Philosophy of Science* 31, pp. 241–54.

1965

Ackermann, Robert, "Discussion: Deductive Scientific Explanation," *Philosophy of Science* 32, pp. 155–67.

Balmuth, Jerome, "Psychoanalytic Explanation," *Mind* 74, pp. 229–35.

Canfield, John, "Teleological Explanation in Biology: A Reply," *British Journal for the Philosophy of Science* 15, pp. 327–31.

Feyerabend, Paul, "Reply to Criticism," in Robert S. Cohen and Marx Wartofsky, eds., *Boston Studies in the Philosophy of Science* 2 (Dordrecht: D. Reidel Publishing Co.), pp. 223–61.

Gorovitz, Samuel, "Causal Judgments and Causal Explanations," *Journal of Philosophy* 62, pp. 695–711.

□ Harman, Gilbert, "Inference to the Best Explanation," *Philosophical Review* 74, pp. 88–95.

• Hempel, Carl G., *Aspects of Scientific Explanation and Other Essays in the Philosophy of Science*. New York: The Free Press.

• ——(a), "Aspects of Scientific Explanation," in *Aspects*.

Hesse, Mary B., "The Explanatory Function of Metaphor," in Yehoshua Bar-Hillel, ed., *Logic, Methodology and Philosophy of Science*, Proceedings of the 1964 International Congress (Amsterdam: North-Holland Publishing Co.), pp. 249–59.

• Kyburg, Henry E., Jr., "Comment," *Philosophy of Science* 32, pp. 147–51.

Lehman, Hugh, "Functional Explanation in Biology," *Philosophy of Science* 32, pp. 1–19.

——(a), "Teleological Explanation in Biology," *British Journal for the Philosophy of Science* 15, p. 327.

Mischel, Theodore, "Concerning Rational Behaviour and Psychoanalytic Explanation," *Mind* 74, pp. 71–78.

Newman, Fred, "Discussion: Explanation Sketches," *Philosophy of Science* 32, pp. 168–72.

Putnam, Hilary, "How Not to Talk about Meaning," in Robert S. Cohen and Marx Wartofsky, eds., *Boston Studies in the Philosophy of Science* 2 (Dordrecht: D. Reidel Publishing Co.), pp. 205–22.

• Salmon, Wesley C., "The Status of Prior Probabilities in Statistical Explanation," *Philosophy of Science* 32, pp. 137–46.

Sellars, Wilfrid, "Scientific Realism or Irenic Realism," in Robert S. Cohen and Marx Wartofsky, eds., *Boston Studies in the Philosophy of Science* 2 (Dordrecht: D. Reidel Publishing Co.), pp. 171–204.

Smart, J. J. C., "Conflicting Views About Explanation," in Robert S. Cohen and Marx Wartofsky, eds., *Boston Studies in the Philosophy of Science* 2 (Dordrecht: D. Reidel Publishing Co.), pp. 151–69.

1966

Ackermann, Robert, and Alfred Stenner, "A Corrected Model of Explanation," *Philosophy of Science* 33, pp. 168–71.

• Bromberger, Sylvain, "Why-Questions," in Robert G. Colodny, ed., *Mind and Cosmos* (Pittsburgh: University of Pittsburgh Press), pp. 86–111.

□ Canfield, John, ed., *Purpose in Nature*. Englewood Cliffs, NJ: Prentice-Hall, Inc.

Canfield, John, and Keith Lehrer, "Discussion: A Note on Prediction and Deduction," *Philosophy of Science* 33, pp. 165–67.

□ Carnap, Rudolf, *Philosophical Foundations of Physics* (edited by Martin Gardner). New York: Basic Books. [Reissued in 1974 as *An Introduction to the Philosophy of Science*.]

Collins, Arthur W., "The Use of Statistics in Explanation," *British Journal for the Philosophy of Science* 17, pp. 127–40.

——(a), "Explanation and Causality," *Mind* 75, pp. 482–85.

Dray, William H., *Philosophical Analysis and History*. New York: Harper & Row.

Feyerabend, Paul, "Article-Review: *The Structure of Science* by Ernest Nagel," *British Journal for the Philosophy of Science* 17, pp. 237–48.

Fine, Arthur I., "Explaining the Behavior of Entities," *Philosophical Review* 75, pp. 496–509.

Gruner, Rolf, "Teleological and Functional Explanation," *Mind* 75, pp. 516–26.

Hempel, Carl G., *Philosophy of Natural Science*. Englewood Cliffs, NJ: Prentice-Hall, Inc. [Unlike other introductory texts in philosophy of science, this book has profoundly influenced the views of scientists, especially behavioral scientists, on the nature of scientific explanation.]

Leach, James, "Discussion: Dray on Rational Explanation," *Philosophy of Science* 33, pp. 61–69.

Lehman, Hugh, "R. E. Merton's Concepts of Function and Functionalism," *Inquiry* 9, pp. 274–83.

Ling, J. F., "Explanation in History," *Mind* 75, pp. 589–91.

Madden, Edward H., "Explanation in Psychoanalysis and History," *Philosophy of Science* 33, pp. 278–86.

Mellor, D. H., "Inexactness and Explanation," *Philosophy of Science* 33, pp. 345–59.

Mischel, Theodore, "Pragmatic Aspects of Explanation," *Philosophy of Science* 33, pp. 40–60.

Scriven, Michael, "Causes, Connections, and Conditions in History," in Dray (1966), pp. 238–64.

Simon, Herbert A., and Nicholas Rescher, "Cause and Counterfactual," *Philosophy of Science* 33, pp. 323–40.

Stegmüller, W., "Explanation, Prediction, Scientific Systematization, and Non-Explanatory Information," *Ratio* 8, pp. 1–24.

1967

Alston, William, "Wants, Actions, and Causal Explanation," in H. Casteneda, ed., *Intentionality, Minds, and Perceptions* (Detroit: Wayne State University Press).

Angel, R. B., "Discussion: Explanation and Prediction: A Plea for Reason," *Philosophy of Science* 34, pp. 276–82.

Coffa, J. Alberto, "Feyerabend on Explanation and Reduction," *Journal of Philosophy* 64, pp. 500–8.

Cunningham, Frank, "More on Understanding in the Social Sciences," *Inquiry* 10, pp. 321–26.

Dietl, Paul, "Paresis and the Alleged Asymmetry Between Explanation and Prediction," *British Journal for the Philosophy of Science* 17, pp. 313–18.

Goh, S. T., "Discussion: Newman and Explanation Sketches," *Philosophy of Science* 34, pp. 273–75.

Gruner, Rolf, "Understanding in the Social Sciences and History," *Inquiry* 10, pp. 151–63.

Kim, Jaegwon, "Explanation in Science," in Paul Edwards, ed., *The Encyclopedia of Philosophy* (New York: Macmillan Publishing Co. and Free Press), vol. 3, pp. 159–63.

Madell, Geoffrey, "Action and Causal Explanation," *Mind* 76, pp. 34–48.

Malcolm, Norman, "Explaining Behavior," *Philosophical Review* 76, pp. 97–104.

Noble, Denis, "Charles Taylor on Teleological Explanation," *Analysis* 27, pp. 96–103.

☐ Salmon, Wesley C., *The Foundations of Scientific Inference*. Pittsburgh: University of Pittsburgh Press.

Shope, Robert K., "Explanation in Terms of 'the Cause'," *Journal of Philosophy* 64, pp. 312–20.

Suchting, W. A., "Deductive Explanation and Prediction Revisited," *Philosophy of Science* 34, pp. 41–52.

Taylor, Charles, "Teleological Explanation—a Reply to Denis Noble," *Analysis* 27, pp. 141–43.

The Third Decade

1968

•Beckner, Morton, *The Biological Way of Thought*. Berkeley & Los Angeles: University of California Press.

Brody, B. A., "Confirmation and Explanation," *Journal of Philosophy* 65, pp. 282–99.

Coffa, J. A., "Discussion: Deductive Predictions," *Philosophy of Science* 35, pp. 279–83.

Finn, D. R., "Categories of Psychological Explanation," *Mind* 77, pp. 550–55.

• Fodor, Jerry A., *Psychological Explanation*. New York: Random House.

——(a), "The Appeal to Tacit Knowledge in Psychological Explanation," *Journal of Philosophy* 65, pp. 627–40.

Good, I. J., "Corroboration, Explanation, Evolving Probability, Simplicity and a Sharpened Razor," *British Journal for the Philosophy of Science* 19, pp. 123–43.

Harman, Gilbert, "Knowledge, Inference, and Explanation," *American Philosophical Quarterly* 5, pp. 161–73.

• Hempel, Carl G., "Maximal Specificity and Lawlikeness in Probabilistic Explanation," *Philosophy of Science* 35, pp. 116–33.

Humphreys, Willard C., "Discussion: Statistical Ambiguity and Maximal Specificity," *Philosophy of Science* 35, pp. 112–15.

Leach, James, "Explanation and Value Neutrality," *British Journal for the Philosophy of Science* 19, pp. 93–108.

Martin, Michael, "Situational Logic and Covering Law Explanation in History," *Inquiry* 11, pp. 388–99.

Massey, Gerald J., "Hempel's Criterion of Maximal Specificity," *Philosophical Studies* 19, pp. 43–47.

Mellor, D. H., "Two Fallacies in Charles Taylor's *Explanation of Behaviour*," *Mind* 77, pp. 124–26.

Paluch, Stanley, "The Covering Law Model of Historical Explanation," *Inquiry* 11, pp. 368–87.

Wright, Larry, "The Case Against Teleological Reductionism," *British Journal for the Philosophy of Science* 19, pp. 211–23.

1969

Aronson, Jerrold L., "Explanations without Laws," *Journal of Philosophy* 66, pp. 541–47.

Boden, Margaret A., "Miracles and Scientific Explanation," *Ratio* 11, pp. 137–44.

Goodfield, June, "Theories and Hypotheses in Biology: Theoretical Entities and Functional Explanation," in Robert S. Cohen and Marx Wartofsky, eds., *Boston Studies in the Philosophy of Science* 5, pp. 421–49.

Gruner, Rolf, "The Notion of Understanding: Replies to Cunningham and Van Evra," *Inquiry* 12, pp. 349–56.

Hanna, Joseph, "Explanation, Prediction, Description, and Information," *Synthese* 20, pp. 308–44.

Hein, Hilde, "Molecular Biology vs. Organicism: The Enduring Dispute between Mechanism and Vitalism," *Synthese* 20, pp. 238–53.

Iseminger, Gary, "Malcolm on Explanations and Causes," *Philosophical Studies* 20, pp. 73–77.

• Jeffrey, Richard C., "Statistical Explanation vs. Statistical Inference," in Nicholas Rescher, ed., *Essays in Honor of Carl G. Hempel* (Dordrecht: D. Reidel Publishing Co.), pp. 104–13. [Reprinted in Salmon, et al. (1971).]

Levi, Isaac, "Are Statistical Hypotheses Covering Laws?" *Synthese* 20, pp. 297–307.

Levison, A. B., and I. Thalberg, "Essential and Causal Explanations," *Mind* 78, pp. 91–101.

Macklin, Ruth, "Explanation and Action: Recent Issues and Controversies," *Synthese* 20, pp. 388–415.

Manier, Edward, "'Fitness' and Some Explanatory Patterns in Biology," *Synthese* 20, pp. 185–205.

McCullagh, C. B., "Narrative and Explanation in History," *Mind* 78, pp. 256–61.

Rescher, Nicholas, "Lawfulness as Mind-Dependent," in Nicholas Rescher, ed., *Essays in Honor of Carl G. Hempel* (Dordrecht: D. Reidel Publishing Co.), pp. 178–97.

Rosenkrantz, Roger D., "On Explanation," *Synthese* 20, pp. 335–70.

☐ Salmon, Wesley C., "Partial Entailment as a Basis for Inductive Logic," in Nicholas Rescher, ed., *Essays in Honor of Carl G. Hempel* (Dordrecht: D. Reidel), pp. 47–82.

Stegmüller, Wolfgang, *Wissenschaftliche Erklärung und Begründung*. Berlin, Heidelberg, & New York: Springer-Verlag.

Van Evra, James W., "Understanding in the Social Sciences Revisited," *Inquiry* 12, pp. 347–49.

Wilson, Fred, "Explanation in Aristotle, Newton, and Toulmin," *Philosophy of Science* 36, part I, pp. 291–310; part II, pp. 400–28.

1970

Ayala, Francisco J., "Teleological Explanations in Evolutionary Biology," *Philosophy of Science* 37, pp. 1–15.

Borger, R., and F. Cioffi, *Explanation in the Behavioural Sciences*. Cambridge: The University Press.

Brittan, Gordon, Jr., "Explanation and Reduction," *Journal of Philosophy* 67, pp. 446–57.

Churchland, Paul, "The Logical Character of Action-Explanations," *Philosophical Review* 79, pp. 214–36.

Ellis, B. D., "Explanation and the Logic of Support," *Australasian Journal of Philosophy* 48, pp. 177–89.

Goh, S. T., "Some Observations on the Deductive-Nomological Theory," *Mind* 79, pp. 408–14.

—— (a), "The Logic of Explanation in Anthropology," *Inquiry* 13, pp. 339–59.

• Greeno, James G., "Evaluation of Statistical Hypotheses Using Information Transmitted," *Philosophy of Science* 37, pp. 279–93. [Reprinted under the title "Explanation and Information" in Salmon, et al. (1971).]

Hedman, C. G., "Gustafson on Explanation in Psychology," *Mind* 79, pp. 272–74.

• Hesse, Mary, *Models and Analogies in Science*. Notre Dame: University of Notre Dame Press.

Kyburg, Henry E., Jr., "Discussion: More on Maximal Specificity," *Philosophy of Science* 37, pp. 295–300.

Margolis, Joseph, "Puzzles regarding Explanation by Reasons and Explanation by Causes," *Journal of Philosophy* 67, pp. 187–95.

Morgan, Charles G., "Discussion: Kim on Deductive Explanation," *Philosophy of Science* 37, pp. 434–39.

Nissen, Lowell, "Canfield's Functional Translation Schema," *British Journal for the Philosophy of Science* 21, pp. 193–95.

Omer, I. A., "On the D-N Model of Scientific Explanation," *Philosophy of Science* 37, pp. 417–33.

☐ Pauling, Linus, *Vitamin C and the Common Cold*. San Francisco: W. H. Freeman and Co.

• Rescher, Nicholas, *Scientific Explanation*. New York: Free Press.

• Salmon, Wesley C., "Statistical Explanation," in Robert G. Colodny, ed., *The Nature and Function of Scientific Theories* (Pittsburgh: University of Pittsburgh Press), pp. 173–231. [Reprinted in Salmon et al. (1971).]

☐ —— (a), ed., *Zeno's Paradoxes*. Indianapolis: Bobbs-Merrill.

Schlegel, Richard, "Statistical Explanation in Physics: The Copenhagen Interpretation," *Synthese* 21, pp. 65–82.

Suppes, Patrick, *A Probabilistic Theory of Causality*. Amsterdam: North-Holland.

Taylor, Charles, "Explaining Action," *Inquiry* 13, pp. 54–89.

1971

Achinstein, Peter, *Law and Explanation*. Oxford: Clarendon Press.

Alexander, Peter, "Psychoanalysis and the Explanation of Behaviour," *Mind* 80, pp. 391–408.

Alston, William P., "The Place of Explanation of Particular Facts in Science," *Philosophy of Science* 38, pp. 13–34.

Economos, John James, "Explanation: What's It All About?" *Australasian Journal of Philosophy* 49, pp. 139–45.

Fetzer, James H., "Dispositional Probabilities," in Roger C. Buck and Robert S. Cohen, eds., *PSA 1970* (Dordrecht: D. Reidel Publishing Co.), pp. 473–82.

Gaukroger, Stephen, *Explanatory Structures*. Brighton: Harvester Press.

Greeno, James G., "Theoretical Entities in Statistical Explanation," in Roger C. Buck and Robert S. Cohen, eds., *PSA 1970* (Dordrecht: D. Reidel Publishing Co.), pp. 3–26.

Hovard, Richard B., "Theoretical Reduction: The Limits and Alternatives to Reductive Methods in Scientific Explanation," *Philosophy of the Social Sciences* 1, pp. 83–100.

Jeffrey, Richard C., "Remarks on Explanatory Power," in Roger C. Buck and Robert S. Cohen, eds., *PSA 1970* (Dordrecht: D. Reidel Publishing Co.), pp. 40–46.

Martin, Michael, "Neurophysiological Reduction and Psychological Explanation," *Philosophy of the Social Sciences* 1, pp. 161–70.

Mullane, Harvey, "Psychoanalytic Explanation and Rationality," *Journal of Philosophy* 68, pp. 413–26.

Nickles, Thomas, "Covering Law Explanation," *Philosophy of Science* 38, pp. 542–61.

Nilson, Sten Sparre, "Covering Laws in Historical Practice," *Inquiry* 14, pp. 445–63.

Ruse, Michael, "Discussion: Functional Statements in Biology," *Philosophy of Science* 38, pp. 87–95.

Salmon, Wesley C., "Explanation and Relevance," in Roger C. Buck and Robert S. Cohen, eds., *PSA 1970* (Dordrecht: D. Reidel Publishing Co.), pp. 27–39.

• Salmon, Wesley C., et al., *Statistical Explanation and Statistical Relevance*. Pittsburgh: University of Pittsburgh Press.

• von Wright, G. H., *Explanation and Understanding*. Ithaca, NY: Cornell University Press.

Watson, Patty Jo, Steven A. LeBlanc, and Charles Redman, *Explanation in Archaeology: An Explicitly Scientific Approach*. New York: Columbia University Press. [Second edition, Watson et al., (1984).]

White, J. E., "Avowed Reasons and Causal Explanation," *Mind* 80, pp. 238–45.

Zaffron, Richard, "Identity, Subsumption, and Scientific Explanation," *Journal of Philosophy* 68, pp. 849–60.

1972

Ball, Terrence, "On 'Historical' Explanation," *Philosophy of the Social Sciences* 2, pp. 181–92.

   ☐ Binford, Lewis, *An Archaeological Perspective*. New York: Harcourt.

Boden, Margaret A., *Purposive Explanation in Psychology*. Cambridge, MA: Harvard University Press.

Brody, Baruch, "Towards an Aristotelian Theory of Scientific Explanation," *Philosophy of Science* 39, pp. 20–31.

Fabian, Robert G., "Human Behavior in Deductive Social Theory: The Example of Economics," *Inquiry* 15, pp. 411–33.

Harré, Rom, and P. F. Secord, *The Explanation of Social Behaviour*. Totowa, N.J.: Rowman and Littlefield.

Lehman, Hugh, "Statistical Explanation," *Philosophy of Science* 39, pp. 500–6.

Martin, Michael, "Confirmation and Explanation," *Analysis* 32, pp. 167–69.

——(a), "Explanation in Social Science: Some Recent Work," *Philosophy of the Social Sciences* 2, pp. 61–81.

Miettinen, Seppo K., "Discussion: On Omer's Model of Explanation," *Philosophy of Science* 39, pp. 249–51.

Morgan, Charles G., "Discussion: On Two Proposed Models of Explanation," *Philosophy of Science* 39, pp. 74–81.

   ☐ Nye, Mary Jo, *Molecular Reality*. London: Macdonald.

Rapoport, Anatol, "Explanatory Power and Explanatory Appeal of Theories," *Synthese* 24, pp. 321–42.

Tuomela, Raimo, "Deductive Explanation of Scientific Laws," *Journal of Philosophical Logic* 1, pp. 369–92.

Varela, Francisco G., and Humberto R. Maturana, "Discussion: Mechanism and Biological Explanation," *Philosophy of Science* 39, pp. 378–82.

Wallace, William A., *Causality and Scientific Explanation*. Ann Arbor: University of Michigan Press.

White, J. E., "Hedman on Explanation," *Mind* 81, pp. 595–96.

Wimsatt, William, "Teleology and the Logical Structure of Function Statements," *Studies in the History and Philosophy of Science* 3, pp. 1–80.

Wright, Larry, "Explanation and Teleology," *Philosophy of Science* 39, pp. 204–18.

——(a), "Discussion: A Comment on Ruse's Analysis of Function Statements," *Philosophy of Science* 39, pp. 512–14.

1973

Bennett, P. W., "Avowed Reasons and the Covering Law Model," *Mind* 82, pp. 606–7.

• Coffa, J. Alberto, *The Foundations of Inductive Explanation*. Doctoral dissertation, University of Pittsburgh.

☐ DeWitt, Bryce S., and Neill Graham, *The Many-Worlds Interpretation of Quantum Mechanics*. Princeton: Princeton University Press.

Finocchiaro, Maurice A., *History of Science as Explanation*. Detroit: Wayne State University Press.

Gower, Barry, "Martin on Confirmation and Explanation," *Analysis* 33, pp. 107–9.

Hull, David, "A Belated Reply to Gruner," *Mind* 82, pp. 437–38.

Krausser, Peter, "A Cybernetic Systemstheoretical Approach to Rational Understanding and Explanation, especially Scientific Revolutions with Radical Meaning Change," *Ratio* 15, pp. 221–46.

☐ Lewis, David, *Counterfactuals*. Cambridge, MA.: Harvard University Press.

Morgan, Charles G., "Discussion: Omer on Scientific Explanation," *Philosophy of Science* 40, pp. 110–17.

Nickles, Thomas, "Discussion: Explanation and Description Relativity," *Philosophy of Science* 40, pp. 408–14.

Ruse, Michael, *Philosophy of Biology*. London: Hutchinson University Library.

——(a), "Discussion: A Reply to Wright's Analysis of Functional Statements," *Philosophy of Science* 40, pp. 277–80.

——(b), "Teleological Explanations and the Animal World," *Mind* 82, pp. 433–36.

Salmon, Wesley C., "Reply to Lehman," *Philosophy of Science* 40, pp. 397–402.

Shelanski, V. B., "Nagel's Translations of Teleological Statements: A Critique," *British Journal for the Philosophy of Science* 24, pp. 397–401.

Sklar, Lawrence, "Statistical Explanation and Ergodic Theory," *Philosophy of Science* 40, pp. 194–212.

Srzednicki, Jan, "Statistical Indeterminism and Scientific Explanation," *Synthese* 26, pp. 197–204.

• Stegmüller, Wolfgang, *Probleme und Resultate der Wissenschaftstheorie und analytischen Philosophie*, Band 4, Studienausgabe Teil E. Berlin/New York: Springer-Verlag.

Steinberg, Danny, "Discussion: Nickles on Intensionality and the Covering Law Model," *Philosophy of Science* 40, pp. 403–7.

Stemmer, Nathan, "Discussion: Brody's Defense of Essentialism," *Philosophy of Science* 40, pp. 393–96.

Tondl, Ladislav, *Scientific Procedures, Boston Studies in the Philosophy of Science* 10, chap. V.

Wright, Larry, "Function," *Philosophical Review* 82, pp. 139–68.

1974

Bogen, James, "Moravcsik on Explanation," *Synthese* 28, pp. 19–26.

Brody, Baruch, "More on Confirmation and Explanation," *Philosophical Studies* 26, pp 73–75.

Buroker, Jill Vance, "Kant, the Dynamical Tradition, and the Role of Matter in Explanation," in Kenneth F. Schaffner and Robert S. Cohen, eds., *PSA 1972* (Dordrecht: D. Reidel Publishing Co.), pp. 153–64.

• Coffa, J. Alberto, "Hempel's Ambiguity," *Synthese* 28, pp. 141–63.

Collins, Paul W., "The Present Status of Anthropology as an Explanatory Science," in Raymond J. Seeger and Robert S. Cohen, eds., *Philosophical Foundations of Physics, Boston Studies in the Philosophy of Science* 11, pp. 337–48.

• Fetzer, James H., "Grünbaum's 'Defense' of the Symmetry Thesis," *Philosophical Studies* 25, pp. 173–87.

• ——(a), "Statistical Explanations," in Kenneth Schaffner and Robert S. Cohen, eds., *PSA 1972* (Dordrecht: D. Reidel Publishing Co.), pp. 337–47.

• ——(b), "A Single Case Propensity Theory of Explanation," *Synthese* 28, pp. 87–97.

• Friedman, Michael, "Explanation and Scientific Understanding," *Journal of Philosophy* 71, pp. 5–19.

• Grene, Marjorie, *The Understanding of Nature: Essays in the Philosophy of Biology*. Dordrecht: D. Reidel Publishing Co.

Hopson, Ronald C., "The Objects of Acceptance: Competing Scientific Explanations," in Kenneth Schaffner and Robert S. Cohen, eds., *PSA 1972* (Dordrecht: D. Reidel Publishing Co.), pp. 349-63.

Klein, Martin J., "Boltzmann, Monocycles and Mechanical Explanation," in Raymond J. Seeger and Robert S. Cohen, eds., *Philosophical Foundations of Physics, Boston Studies in the Philosophy of Science* 11, pp. 155-75.

☐ Mackie, J. L., *The Cement of the Universe*. Oxford: Clarendon Press.

Mayr, Ernst, "Teleologica and Teleonomic, a New Analysis," in Robert S. Cohen and Marx Wartofsky, eds., *Boston Studies in the Philosophy of Science* 14, pp. 91-118.

Moravcsik, Julius M. E., "Aristotle on Adequate Explanations," *Synthese* 28, pp. 3-18.

Salmon, Wesley C., "Comments on 'Hempel's Ambiguity' by J. Alberto Coffa," *Synthese* 28, pp. 165-69.

Singleton, J., "The Explanatory Power of Chomsky's Transformational Generative Grammar," *Mind* 83, pp. 429-31.

• Teller, Paul, "On Why-Questions," *Nous* 8, pp. 371-80.

Thorpe, Dale A., "Discussion: The Quartercentenary Model of D-N Explanation," *Philosophy of Science* 41, pp. 188-95.

Wigner, Eugene P., "Physics and the Explanation of Life," in Raymond J. Seeger and Robert S. Cohen, eds., *Philosophical Foundations of Physics, Boston Studies in the Philosophy of Science* 11, pp. 119-32.

Wright, Larry, "Mechanisms and Purposive Behavior," *Philosophy of Science* 41, pp. 345-60.

1975

Achinstein, Peter, "The Object of Explanation," in Körner (1975), pp. 1-45. [See also comments by Rom Harré and Mary Hesse, and Achinstein's reply.]

Baublys, Kenneth K., "Discussion: Comments on Some Recent Analyses of Functional Statements in Biology," *Philosophy of Science* 42, pp. 469-86.

• Brody, Baruch, "The Reduction of Teleological Sciences," *American Philosophical Quarterly* 12, pp. 69-76.

Burian, Richard M., "Conceptual Change, Cross-Theoretical Explanation, and the Unity of Science," *Synthese* 32, pp. 1-28.

Crowell, E., "Causal Explanation and Human Action," *Mind* 84, pp. 440-42.

Cummins, Robert, "Functional Analysis," *Journal of Philosophy* 72, 741-65.

Eglin, Peter, "What Should Sociology Explain – Regularities, Rules or Interpretations?" *Philosophy of the Social Sciences* 5, pp. 277-92.

Englehardt, H. T., Jr., and S. F. Spicker, eds., *Evaluation and Explanation in the Biomedical Sciences*. Dordrecht: D. Reidel Publishing Co.

Fetzer, James H., "On the Historical Explanation of Unique Events," *Theory and Decision* 6, pp. 87-97.

——(a), "Discussion Review: Achinstein's *Law and Explanation*," *Philosophy of Science* 42, pp. 320-33.

☐ Few, A. A., "Thunder," *Scientific American* 233, no. 1 (July), pp. 80-90.

Good, I. J., "Explicativity, Corroboration, and the Relative Odds of Hypotheses," *Synthese* 30, pp. 39-74.

Koertge, Noretta, "An Exploration of Salmon's S-R Model of Explanation," *Philosophy of Science* 42, pp. 270-74.

——(a), "Popper's Metaphysical Research Program for the Human Sciences," *Inquiry* 18, pp. 437-62.

• Körner, Stephan, ed., *Explanation*. Oxford: Basil Blackwell.

Martin, Michael, "Explanation and Confirmation Again," *Analysis* 36, pp. 41-42.

Minton, Arthur J., "Discussion: Wright and Taylor: Empiricist Teleology," *Philosophy of Science* 42, pp. 299-306.

Mucciolo, Laurence F., "Neurophysiological Reduction, Psychological Explanation, and Neuropsychology," *Philosophy of the Social Sciences* 5, pp. 451-62.

Salmon, Wesley C., "Theoretical Explanation," in Körner (1975), pp. 118–45. [See also comments by L. J. Cohen and D. H. Mellor, and Salmon's reply.]

☐ ——(a), *Space, Time, and Motion: A Philosophical Introduction*. Encino, CA: Dickenson Publishing Co. [Second edition, Minneapolis: University of Minnesota Press, 1980.]

Sanders, Robert E., and Larry W. Martin, "Grammatical Rules and Explanation of Behavior," *Inquiry* 18, pp. 65–82.

• Scriven, Michael, "Causation as Explanation," *Nous* 9, pp. 3–16.

Studdert-Kennedy, Gerald, *Evidence and Explanation in Social Science*. London: Routledge & Kegan Paul.

1976

Addis, Laird, "On Defending the Covering Law Model," in R. S. Cohen, C. A. Hooker, A. C. Michalos, and J. W. Van Evra, eds., *PSA 1974* (Dordrecht: D. Reidel Publishing Co.), pp. 361–68.

Beckner, Morton, "Functions and Teleology," in Grene and Mendelsohn (1976), pp. 197–212.

☐ Belnap, Nuel D., Jr., and J. B. Steel, Jr., *The Logic of Questions and Answers*. New Haven: Yale University Press.

Boorse, Christopher, "Wright on Functions," *Philosophical Review* 85, pp. 70–86.

Cherry, Christopher, "Explanation and Explanation by Hypothesis," *Synthese* 33, pp. 315–40.

Davidson, Donald, "Hempel on Explaining Action," *Erkenntnis* 10, pp. 239–54.

Downes, Chauncey, "Functional Explanations and Intentions," *Philosophy of the Social Sciences* 6, pp. 215–25.

Fetzer, James H., "The Likeness of Lawlikeness," in Robert S. Cohen et al., eds., *Boston Studies in the Philosophy of Science* 32 (Dordrecht: D. Reidel Publishing Co.), pp. 337–91.

Gardenfors, Peter, "Relevance and Redundancy in Deductive Explanations," *Philosophy of Science* 43, pp. 420–31.

Girill, T. R., "The Problem of Micro-Explanation," in Frederick Suppe and Peter D. Asquith, eds., *PSA 1976*, vol. 1 (East Lansing, MI: Philosophy of Science Assn.), pp. 47–55.

——(a), "Evaluating Micro-Explanations," *Erkenntnis* 10, pp. 387–406.

• Grene, Marjorie, and Everett Mendelsohn, eds., *Topics in the Philosophy of Biology, Boston Studies in the Philosophy of Science* 27, part III, Problems of Explanation in Biology, pp. 145–263.

Grobstein, Clifford, "Organizational Levels and Explanation," in Grene and Mendelsohn (1976), pp. 145–52.

Hempel, C. G., "Dispositional Explanation and the Covering Law Model: Response to Laird Addis," in R. S. Cohen, C. A. Hooker, A. C. Michalos, and J. W. Van Evra, eds., *PSA 1974* (Dordrecht: D. Reidel Publishing Co.), pp. 369–76.

Hesslow, Germund, "Two Notes on the Probabilistic Approach to Causality," *Philosophy of Science* 43, pp. 290–92.

Jobe, Evan, "A Puzzle Concerning D-N Explanation," *Philosophy of Science* 43, pp. 542–49.

Kauffman, Stuart, "Articulation of Parts Explanation in Biology and the Rational Search for Them," in Grene and Mendelsohn (1976), pp. 245–63.

King, John L., "Statistical Relevance and Explanatory Classification," *Philosophical Studies* 30, pp. 313–21.

• Kitcher, Philip, "Explanation, Conjunction, and Unification," *Journal of Philosophy* 73, pp. 207–12.

Krüger, Lorenz, "Discussion: Are Statistical Explanations Possible?" *Philosophy of Science* 43, pp. 129–46.

Levin, Michael D., "The Extensionality of Causation and Causal-Explanatory Contexts," *Philosophy of Science* 43, pp. 266–77.

——(a), "On the Ascription of Functions to Objects, with Special Reference to Inference in Archaeology," *Philosophy of the Social Sciences* 6, pp. 227–34.

• Manninen, J., and R. Tuomela, eds., *Essays on Explanation and Understanding*. Dordrecht: D. Reidel Publishing Co. [A collection of essays pertaining to von Wright (1971).]

Martin, Rex, "Explanation and Understanding in History," in Manninen and Tuomela (1976), pp. 305–34.

McLachlan, Hugh V., "Functionalism, Causation and Explanation," *Philosophy of the Social Sciences* 6, pp. 235–40.

• Mellor, D. H., "Probable Explanation," *Australasian Journal of Philosophy* 54, pp. 231–41.

Morgan, Charles G., "Tuomela on Deductive Explanation," *Journal of Philosophical Logic* 5, pp. 511–25.

Munch, Peter A., "The Concept of Function and Functional Analysis in Sociology," *Philosophy of the Social Sciences* 6, pp. 193–214.

Nathan, N. M. L., "On the Non-Causal Explanation of Human Action," *Philosophy of the Social Sciences* 6, pp. 241–43.

Niiniluoto, Ilkka, "Inductive Explanation, Propensity, and Action," in Manninen and Tuomela (1976), pp. 335–68.

Pattee, H. H., "Physical Theories of Biological Co-ordination," in Grene and Mendelsohn (1976), pp. 153–73.

☐ Reichenbach, Hans, *Laws, Modalities, and Counterfactuals*. Berkeley & Los Angeles: University of California Press. [Reprint of Reichenbach (1954) with a new foreword by Wesley C. Salmon.]

Riedel, Manfred, "Causal and Historical Explanation," in Manninen and Tuomela (1976), pp. 3–26.

Ringen, Jon D., "Explanation, Teleology, and Operant Behaviorism: A Study of the Experimental Analysis of Purposive Behavior," *Philosophy of Science* 43, pp. 223–53.

☐ Salmon, Wesley C., foreword to Reichenbach (1954).

Simon, Thomas W., "A Cybernetic Analysis of Goal-Directedness," in Frederick Suppe and Peter D. Asquith, eds., *PSA 1976*, vol. 1 (East Lansing, MI: Philosophy of Science Assn.), pp. 56–67.

Tuomela, Raimo, "Causes and Deductive Explanation," in R. S. Cohen, C. A. Hooker, A. C. Michalos, and J. W. Van Evra, eds., *PSA 1974* (Dordrecht: D. Reidel Publishing Co.), pp. 325–60.

——(a), "Explanation and Understanding of Human Behavior," in Manninen and Tuomela (1976), pp. 183–205.

——(b), "Morgan on Deductive Explanation: A Rejoinder," *Journal of Philosophical Logic* 5, pp. 527–43.

van Fraassen, Bas C., "To Save the Phenomena," *Journal of Philosophy* 73, pp. 623–32.

Williams, Mary B., "The Logical Structure of Functional Explanations in Biology," in Frederick Suppe and Peter D. Asquith, eds., *PSA 1976*, vol. 1 (East Lansing, MI: Philosophy of Science Assn.), pp. 37–46.

Wimsatt, William C., "Complexity and Organization," in Grene and Mendelsohn (1976), pp. 174–93.

• Woodfield, Andrew, *Teleology*. New York: Cambridge University Press.

• Wright, Larry, *Teleological Explanations*. Berkeley/Los Angeles/London: University of California Press.

——(a), "Functions," in Grene and Mendelsohn (1976), pp. 213–42.

1977

Achinstein, Peter, "What Is an Explanation?" *American Philosophical Quarterly* 14, pp. 1–16.

——(a), "Function Statements," *Philosophy of Science* 44, pp. 341–67.

Causey, Robert L., *Unity of Science*. Dordrecht: D. Reidel Publishing Co.

• Coffa, J. Alberto, "Probabilities: Reasonable or True?" *Philosophy of Science* 44, pp. 186–98.

Cummins, Robert, "Programs in the Explanation of Behavior," *Philosophy of Science* 44, pp. 269–87.

Cupples, B., "Three Types of Explanation," *Philosophy of Science* 44, pp. 387–408.

Fetzer, James H., "A World of Dispositions," *Synthese* 34, pp. 397–421.

Gluck, Peter, and Michael Schmid, "The Rationality Principle and Action Explanations: Koertge's Reconstruction of Popper's Logic of Action Explanations," *Inquiry* 20, pp. 72–81.

Good, I. J., "Explicativity: A Mathematical Theory of Explanation with Statistical Applications," *Proceedings of the Royal Society of London* 354, pp. 303–30.

Goode, Terry M., "Explanation, Expansion, and the Aims of Historians: Toward an Alternative Account of Historical Explanation," *Philosophy of the Social Sciences* 7, pp. 367–84.

• Hempel, Carl G., "Nachwort 1976: Neuere Ideen zu den Problemen der statistischen Erklärung," in Carl G. Hempel, *Aspekte wissenschaftlicher Erklärung* (Berlin/New York: Walter de Gruyter), pp. 98–123.

Kung, Joan, "Aristotle on Essence and Explanation," *Philosophical Studies* 31, pp. 361–83.

Martin, Michael, "Neurophysiological Reduction and Type Identity," *Philosophy of the Social Sciences* 7, pp. 91–94.

McCarthy, Tim, "On an Aristotelian Model of Scientific Explanation," *Philosophy of Science* 44, pp. 159–66.

Nagel, Ernest, "Teleology Revisited," *Journal of Philosophy* 74, pp. 261–301. [Reprinted in Nagel, 1979.]

Nickles, Thomas, "On the Independence of Singular Causal Explanation in Social Science: Archaeology," *Philosophy of the Social Sciences* 7, pp. 163–87.

—— (a), "Davidson on Explanation," *Philosophical Studies* 31, pp. 141–45.

Rosenberg, Alexander, "Concrete Occurrences vs. Explanatory Facts," *Philosophical Studies* 31, pp. 133–40.

• Salmon, Wesley C., "An 'At-At' Theory of Causal Influence," *Philosophy of Science* 44, pp. 215–24.

—— (a), "Hempel's Conception of Inductive Inference in Inductive-Statistical Explanation," *Philosophy of Science* 44, pp. 180–85.

• —— (b), "Objectively Homogeneous Reference Classes," *Synthese* 36, pp. 399–414.

• —— (c), "A Third Dogma of Empiricism," in Robert Butts and Jaakko Hintikka, eds., *Basic Problems in Methodology and Linguistics* (Dordrecht: D. Reidel Publishing Co.), pp. 149–66.

—— (d), "Indeterminism and Epistemic Relativization," *Philosophy of Science* 44, pp. 199–202.

Sayre, Kenneth, "Statistical Models of Causal Relations," *Philosophy of Science* 44, pp. 203–14.

Shrader, Douglas W., Jr., "Discussion: Causation, Explanation, and Statistical Relevance," *Philosophy of Science* 44, pp. 136–45.

Tuomela, Raimo, *Human Action and its Explanation*. Dordrecht: D. Reidel Publishing Co.

—— (a), "Dispositions, Realism, and Explanation," *Synthese* 34, pp. 457–78.

Utz, Stephen, "Discussion: On Teleology and Organisms," *Philosophy of Science* 44, pp. 313–20.

• van Fraassen, Bas C., "The Pragmatics of Explanation," *American Philosophical Quarterly* 14, pp. 143–50.

von Bretzel, Philip, "Concerning a Probabilistic Theory of Causation Adequate for the Causal Theory of Time," *Synthese* 35, pp. 173–90.

Wright, Larry, "Rejoinder to Utz," *Philosophy of Science* 44, pp. 321–25.

The Fourth Decade

1978

Aronovitch, Hilliard, "Social Explanation and Rational Motivation," *American Philosophical Quarterly* 15, pp. 197–204.

Bridgstock, Martin, and Michael Hyland, "The Nature of Individualistic Explanation: A Further Analysis of Reduction," *Philosophy of the Social Sciences* 8, pp. 265–70.

Cummins, Robert, "Explanation and Subsumption," in Peter D. Asquith and Ian Hacking, eds., *PSA 1978*, vol. 1 (East Lansing, MI: Philosophy of Science Assn.), pp. 163–75.

Derden, J. K., "Reasons, Causes, and Empathetic Understanding," in Peter D. Asquith and Ian Hacking, eds., *PSA 1978*, vol. 1 (East Lansing, MI: Philosophy of Science Assn.), pp. 176–85.

Dorling, Jon, "Discussion: On Explanations in Physics: Sketch of an Alternative to Hempel's Account of the Explanation of Laws," *Philosophy of Science* 45, pp. 136–40.

Essler, Wilhelm K., "A Note on Functional Explanation," *Erkenntnis* 13, pp. 371–76.

Girill, T. R., "Approximative Explanation," in Peter D. Asquith and Ian Hacking, eds., *PSA 1978*, vol. 1 (East Lansing, MI: Philosophy of Science Assn.), pp. 186–96.

Glymour, Clark, "Two Flagpoles Are More Paradoxical Than One," *Philosophy of Science* 45, pp. 118–19.

• Hanna, Joseph, "On Transmitted Information as a Measure of Explanatory Power," *Philosophy of Science* 45, pp. 531–62.

Hempel, Carl G., "Dispositional Explanation," in Raimo Tuomela, ed., *Dispositions* (Dordrecht: D. Reidel Publishing Co.), pp. 137–46. [An extensively revised version of Hempel (1965), pp. 457–63.]

Levin, Michael E., and Margarita Rosa Levin, "Flagpoles, Shadows and Deductive Explanation," *Philosophical Studies* 32, pp. 293–99.

• McMullin, Ernan, "Structural Explanation," *American Philosophical Quarterly* 15, pp. 139–47.

Moor, James H., "Explaining Computer Behavior," *Philosophical Studies* 34, pp. 325–27.

Pinkard, Terry, "Historical Explanation and the Grammar of Theories," *Philosophy of the Social Sciences* 8, pp. 227–40.

Quay, Paul M., "A Philosophical Explanation of the Explanatory Functions of Ergodic Theory," *Philosophy of Science* 45, pp. 47–59.

• Railton, Peter, "A Deductive-Nomological Model of Probabilistic Explanation," *Philosophy of Science* 45, pp. 206–26.

Rosen, Deborah, "In Defense of a Probabilistic Theory of Causality," *Philosophy of Science* 45, pp. 604–13.

Saliers, Don E., "Explanation and Understanding in the Social Sciences," *Philosophy of the Social Sciences* 8, pp. 367–71.

• Salmon, Wesley C., "Why Ask, 'Why?'? – An Inquiry Concerning Scientific Explanation," *Proceedings and Addresses of the American Philosophical Association* 51, pp. 683–705.

Sayre, Kenneth, "Discussion: Masking and Causal Relatedness: An Elucidation," *Philosophy of Science* 45, pp. 633–37.

Shrader, Douglas, Jr., "Discussion: Sayre's Statistical Model of Causal Relations," *Philosophy of Science* 45, pp. 630–32.

Stern, Cindy, "Discussion: On the Alleged Extensionality of 'Causal-Explanatory Contexts'," *Philosophy of Science* 45, pp. 614–25.

Stiner, Mark, "Mathematical Explanation," *Philosophical Studies* 34, pp. 135–51.

Thagard, Paul R., "The Best Explanation: Criteria for Theory Choice," *Journal of Philosophy* 75, pp. 76–92.

1979

Chopra, Y. N., " 'Explaining and Characterizing Human Action," *Mind* 88, pp. 321–33.

Cupples, Brian, "Discussion: Moor and Schlesinger on Explanation," *Philosophy of Science* 46, pp. 645–50.

Enc, Berent, "Function Attributions and Functional Explanations," *Philosophy of Science* 46, pp. 343–65.

Fair, David, "Causation and the Flow of Energy," *Erkenntnis* 14, pp. 219–50.

• Fetzer, James H., and Donald E. Nute, "Syntax, Semantics, and Ontology: A Probabilistic Causal Calculus," *Synthese* 40, pp. 453–95.

☐ Gardner, Michael, "Realism and Instrumentalism in the 19th Century," *Philosophy of Science* 46, pp. 1–34.

Hanna, Joseph F., "An Interpretive Survey of Recent Research on Scientific Explanation," in Peter D. Asquith and Henry E. Kyburg, Jr., *Current Research in Philosophy of Science* (East Lansing, MI: Philosophy of Science Assn.), pp. 291–316.

Martin, Michael, "Reduction and Typical Individuals Again," *Philosophy of the Social Sciences* 9, pp. 77–79.

Meixner, John B., "Homogeneity and Explanatory Depth," *Philosophy of Science* 46, pp. 366–81.

Nagel, Ernest, "Teleology Revisited," in *Teleology Revisited and Other Essays in the Philosophy and History of Science* (New York: Columbia University Press), pp. 275–316. [First published in 1977.]

Nerlich, Graham, "What Can Geometry Explain," *British Journal for the Philosophy of Science* 30, pp. 69–83.

Salmon, Wesley C., "Postscript: Laws in Deductive-Nomological Explanation—An Application of the Theory of Nomological Statements," in Wesley C. Salmon, ed., *Hans Reichenbach: Logical Empiricist* (Dordrecht: D. Reidel Publishing Co.), pp. 691–94.

☐ (a), "Propensities: A Discussion Review," *Erkenntnis* 14, pp. 183–216.

Thomas, J/P, "Homogeneity Conditions on the Statistical Relevance Model of Explanation," *Philosophical Studies* 36, pp. 101–6.

Van Parijs, Philippe, "Functional Explanation and the Linguistic Analogy," *Philosophy of the Social Sciences* 9, pp. 425–44.

Wilson, Ian, "Explanatory and Inferential Conditionals," *Philosophical Studies* 35, pp. 269–78.

Wolfson, Paul, and James Woodward, "Scientific Explanation and Sklar's Views of Space and Time," *Philosophy of Science* 46, pp. 287–94.

Woodward, James, "Scientific Explanation," *British Journal for the Philosophy of Science* 30, pp. 41–67.

1980

Cooke, Roger, "Discussion: A Trivialization of Nagel's Definition of Explanation for Statistical Laws," *Philosophy of Science* 47, pp. 644–45.

Cupples, B., "Four Types of Explanation," *Philosophy of Science* 47, pp. 626–29.

Currie, Gregory, "The Role of Normative Assumptions in Historical Explanation," *Philosophy of Science* 47, pp. 456–73.

Fetzer, James H., and Donald E. Nute, "A Probabilistic Causal Calculus: Conflicting Conceptions," *Synthese* 44, pp. 241–46.

Forge, John, "The Structure of Physical Explanation," *Philosophy of Science* 47, pp. 203–26.

☐ Frohlich, Cliff, "The Physics of Somersaulting and Twisting," *Scientific American* 242 (March), pp. 154–64.

Gardenfors, Peter, "A Pragmatic Approach to Explanations," *Philosophy of Science* 47, pp. 404–23.

Good, I. J., "Some Comments on Probabilistic Causality," *Pacific Philosophical Quarterly* 61, pp. 301–4.

—— (a), "A Further Note on Probabilistic Causality: Mending the Chain," *Pacific Philosophical Quarterly* 61, pp. 452–54.

Gruender, David, "Scientific Explanation and Norms in Science," in Peter D. Asquith and Ronald N. Giere, eds., *PSA 1980*, vol. 1 (East Lansing, MI: Philosophy of Science Assn.), pp. 329–35.

Hooker, C. A., "Explanation, Generality, and Understanding," *Australasian Journal of Philosophy* 58, pp. 284–90.

Humphreys, Paul, "Cutting the Causal Chain," *Pacific Philosophical Quarterly* 61, pp. 305–14.

Klein, Barbara V. E., "What Should We Expect of a Theory of Explanation," in Peter D. Asquith and Ronald N. Giere, eds., *PSA 1980*, vol. 1 (East Lansing, MI: Philosophy of Science Assn.), pp. 319–28.

Kline, A. David, "Screening-off and the Temporal Asymmetry of Explanation," *Analysis* 40, pp. 139–43.

Lambert, Karel, "Explanation and Understanding: An Open Question?" in Risto Hilpinin, ed., *Rationality in Science* (Dordrecht: D. Reidel Publishing Co.), pp. 29–34.

Laymon, Ronald, "Idealization, Explanation, and Confirmation," in Peter D. Asquith and Ronald N. Giere, eds., *PSA 1980*, vol. 1 (East Lansing, MI: Philosophy of Science Assn.), pp. 336–50.

Omer, I. A., "Minimal Law Explanation," *Ratio* 22, pp. 155–66.

Post, John F., "Infinite Regresses of Justification and of Explanation," *Philosophical Studies* 38, pp. 31–52.

• Railton, Peter, *Explaining Explanation*. Ph.D. dissertation, Princeton University.

☐ Salmon, Wesley C., "Probabilistic Causality," *Pacific Philosophical Quarterly* 61, pp. 50–74.

☐ Skyrms, Brian, *Causal Necessity*. New Haven: Yale University Press.
• van Fraassen, Bas C., *The Scientific Image*. Oxford: Clarendon Press.
Woodward, James, "Developmental Explanation," *Synthese* 44, pp. 443–66.

1981

Addis, Laird, "Dispositions, Explanation, and Behavior," *Inquiry* 24, pp. 205–27.
Cooke, Roger M., "Discussion: A Paradox in Hempel's Criterion of Maximal Specificity," *Philosophy of Science* 48, pp. 327–28.
Creary, Lewis G., "Causal Explanation and the Reality of Natural Component Forces," *Pacific Philosophical Quarterly* 62, pp. 148–57.
• Fetzer, James H., *Scientific Knowledge*. Dordrecht: D. Reidel Publishing Co.
—— (a), "Probability and Explanation," *Synthese* 48, pp. 371–408.
Gale, George, "The Anthropic Principle," *Scientific American* 245, no. 6 (Dec.), pp. 154–71.
• Garfinkel, Alan, *Forms of Explanation*. New Haven and London: Yale University Press.
☐ Hacking, Ian, "Do We See Through a Microscope?" *Pacific Philosophical Quarterly* 62, pp. 305–22.
Hanna, Joseph, "Single Case Propensities and the Explanation of Particular Events," *Synthese* 48, pp. 409–36.
Huff, Douglas, and Stephen Turner, "Rationalizations and the Application of Causal Explanations of Human Action," *American Philosophical Quarterly* 18, pp. 213–20.
• Humphreys, Paul, "Aleatory Explanation," *Synthese* 48, pp. 225–32.
• Kitcher, Philip, "Explanatory Unification," *Philosophy of Science* 48, pp. 507–31.
Matthews, Robert J., "Explaining Explanation," *American Philosophical Quarterly* 18, pp. 71–77.
Niiniluoto, Ilkka, "Statistical Explanation," in G. Floistad, ed., *Contemporary Philosophy 1966–1978* (The Hague: Martinus Nijhoff), pp. 157–87.
—— (a), "Statistical Explanation Reconsidered," *Synthese* 48, pp. 437–72.
Otte, Richard, "A Critique of Suppes' Theory of Probabilistic Causality," *Synthese* 48, pp. 167–90.
Pitt, Joseph C., *Pictures, Images, and Conceptual Change*. Dordrecht: D. Reidel Publishing Co.
• Railton, Peter, "Probability, Explanation, and Information," *Synthese* 48, pp. 233–56.
• Rogers, Ben, "Probabilistic Causality, Explanation, and Detection," *Synthese* 48, pp. 201–23.
Salmon, Merrilee H., "Ascribing Functions to Archaeological Objects," *Philosophy of the Social Sciences* 11, pp. 19–26.
Salmon, Wesley C., "Causality: Production and Propagation," in Peter D. Asquith and Ronald N. Giere, eds., *PSA 1980*, vol. 2 (East Lansing, MI: Philosophy of Science Assn.), pp. 49–69.
☐ —— (a), "Rational Prediction," *British Journal for the Philosophy of Science* 32, pp. 115–25.
Suppes, Patrick, and Mario Zanotti, "When Are Probabilistic Explanations Possible?" *Synthese* 48, pp. 191–200.
Tuomela, Raimo, "Inductive Explanation," *Synthese* 48, pp. 257–94.
Wicken, Jeffrey S., "Causal Explanations in Classical and Statistical Thermodynamics," *Philosophy of Science* 48, pp. 65–77.

1982

Abel, Ruben, "What is an Explanandum?" *Pacific Philosophical Quarterly* 63, pp. 86–92.
Brown, James Robert, "Realism, Miracles, and the Common Cause," in Peter D. Asquith and Thomas Nickles, eds., *PSA 1982*, vol. 1 (East Lansing, MI: Philosophy of Science Assn.), pp. 98–105.
Cohen, G. A., "Functional Explanation, Consequence Explanation, and Marxism," *Inquiry* 25, pp. 27–56.
Ellett, Frederick S., Jr., and David P. Ericson, "On Reichenbach's Principle of the Common Cause," *Pacific Philosophical Quarterly* 64, pp. 330–40.
Farr, James, "Humean Explanations in the Moral Sciences," *Inquiry* 25, pp. 57–80.
Forge, John, "Physical Explanation: With Reference to the Theories of Hempel and Salmon," in Robert

McLaughlin, ed., *What? Where? When? Why?* (Dordrecht: D. Reidel Publishing Co.), pp. 211–29.

Glymour, Clark, "Causal Inference and Causal Explanation," in Robert McLaughlin, ed., *What? Where? When? Why?* (Dordrecht: D. Reidel Publishing Co.), pp. 179–91.

Hutchison, Keith, "What Happened to Occult Qualities in the Scientific Revolution?" *Isis* 73, pp. 233–53.

Leslie, John, "Anthropic Principle, World Ensemble, Design," *American Philosophical Quarterly* 19, pp. 141–52.

Levy, Edwin, "Critical Discussion: Causal-Relevance Explanations: Salmon's Theory and its Relation to Reichenbach," *Synthese* 50, pp. 423–45.

Lipton, Peter, "Nagel Revisited" (Review of Ernest Nagel: *Teleology Revisited*), *British Journal for the Philosophy of Science* 33, pp. 186–94.

MacKinnon, Edward M., *Scientific Explanation and Atomic Physics*. Chicago: University of Chicago Press.

Mandelbaum, Maurice, "G. A. Cohen's Defense of Functional Explanation," *Philosophy of the Social Sciences* 12, pp. 285–88.

Meixner, John, "Are Statistical Explanations Really Explanatory?" *Philosophical Studies* 42, pp. 201–7.

Pork, Andrus, "A Note on Schemes of Historical Explanation: Problem of Status," *Philosophy of the Social Sciences* 12, pp. 409–14.

• Salmon, Merrilee H., *Philosophy and Archaeology*. New York: Academic Press.

Salmon, Wesley C., "Comets, Pollen, and Dreams: Some Reflections on Scientific Explanation," in Robert McLaughlin, ed., *What? Where? When? Why?* (Dordrecht: D. Reidel Publishing Co.), pp. 155–78.

——(a), "Further Reflections," ibid., pp. 231–80.

Taylor, Denise Meyerson, "Actions, Reasons and Causal Explanation," *Analysis* 42, pp. 216–19.

van Fraassen, Bas C., "Rational Belief and the Common Cause Principle," in Robert McLaughlin, ed., *What? Where? When? Why?* (Dordrecht: D. Reidel Publishing Co.), pp. 193–209.

1983

• Achinstein, Peter, *The Nature of Explanation*. New York: Oxford University Press.

Cartwright, Nancy, *How the Laws of Physics Lie*. New York: Oxford University Press.

• Cummins, Robert, *The Nature of Psychological Explanation*. Cambridge, MA: MIT Press.

Fetzer, James H., "Probabilistic Explanations," *PSA 1982*, vol. 2, pp. 194–207.

• Hanna, Joseph, "Probabilistic Explanation and Probabilistic Causality," in Peter D. Asquith and Thomas Nickles, eds., *PSA 1982*, vol. 2, pp. 181–93.

Hausman, Daniel M., "Causal and Explanatory Asymmetry," in Peter D. Asquith and Thomas Nickles, eds., *PSA 1982*, vol. 2, pp. 43–54.

Howson, Colin, "Statistical Explanation and Statistical Support," *Erkenntnis* 20, pp. 61–78.

• Humphreys, Paul, "Aleatory Explanation Expanded," in Peter Asquith and Thomas Nickles, eds., *PSA 1982*, vol. 2, pp. 208–23.

Perry, Clifton, "The Explanatory Efficacy of Individualism," *Philosophy of the Social Sciences* 13, pp. 65–68.

Platts, Mark, "Explanatory Kinds," *British Journal for the Philosophy of Science* 34, pp. 113–48.

Short, T. L., "Teleology in Nature," *American Philosophical Quarterly* 20, pp. 311–20.

Sober, Elliott, "Equilibrium Explanation," *Philosophical Studies* 43, pp. 201–10.

Walt, Stephen, "A Note on Mandelbaum's 'G. A. Cohen's Defense of Functional Explanation'," *Philosophy of the Social Sciences* 13, pp. 483–86.

1984

Achinstein, Peter, "A Type of Non-Causal Explanation," in Peter A. French, Theodore E. Uehling, Jr., and Howard K. Wetterstein, eds., *Causation and Causal Theories, Midwest Studies in Philosophy* 9, (Minneapolis: University of Minnesota Press) pp. 221–44.

Amundson, Ron, and Laurence Smith, "Clark Hull, Robert Cummins, and Functional Analysis," *Philosophy of Science* 51, pp. 657–60.

Collins, Arthur, "Action, Causality, and Teleological Explanation," in Peter A. French, Theodore E. Uehling, Jr., and Howard K. Wetterstein, eds., *Causation and Causal Theories, Midwest Studies in Philosophy* 9, pp. 345–70.

Fetzer, James H., "Review of Peter Achinstein, *The Nature of Explanation*," *Philosophy of Science* 51, pp. 516–19.

☐ Leplin, Jerrett, ed., *Scientific Realism*. Berkeley/Los Angeles/London: University of California Press.

McMullin, Ernan, "Two Ideals of Explanation in Natural Science," in Peter A. French, Theodore E. Uehling, Jr., and Howard K. Wetterstein, eds., *Causation and Causal Theories, Midwest Studies in Philosophy* 9, pp. 205–20.

Nissen, Lowell, "Discussion: Woodfield's Analysis of Teleology," *Philosophy of Science* 51, pp. 488–94.

Papineau, David, "Representation and Explanation," *Philosophy of Science* 51, pp. 550–72.

• Salmon, Wesley C., *Scientific Explanation and the Causal Structure of the World*. Princeton: Princeton University Press.

☐ ——(a), *Logic*, third edition. Englewood Cliffs, NJ: Prentice-Hall.

• Sintonen, Matti, *The Pragmatics of Scientific Explanation. Acta Philosophica Fennica* 37. Helsinki: Societas Philosophica Fennica.

——(a), "On the Logic of Why-Questions," in Peter D. Asquith and Philip Kitcher, eds., *PSA 1984*, vol. 1 (East Lansing, MI: Philosophy of Science Assn.), pp. 168–76.

• Sober, Elliott, "Common Cause Explanation," *Philosophy of Science* 51, pp. 212–41.

Tuomela, Raimo, "Social Action-Functions," *Philosophy of the Social Sciences* 14, pp. 133–48.

• Watkins, John, *Science and Scepticism*. Princeton: Princeton University Press.

Watson, Patty Jo, Steven A. LeBlanc, and Charles Redman, *Archaeological Explanation: The Scientific Method in Archaeology*. New York: Columbia University Press.

Woodward, James, "Explanatory Asymmetries," *Philosophy of Science* 51, pp. 421–42.

——(a), "A Theory of Singular Causal Explanation," *Erkenntnis* 21, pp. 231–62.

1985

Apel, Karl-Otto, *Understanding and Explanation: A Transcendental-Pragmatic Perspective*. Cambridge, MA: MIT Press.

Carleton, Lawrence R., "Levels in Discription and Explanation," *Philosophy Research Archives* 9, pp. 89–110.

☐ Churchland, Paul M., and Clifford A. Hooker, *Images of Science*. Chicago: University of Chicago Press.

Doppelt, Gerald, "Finocchiaro on Rational Explanation," *Synthese* 62, pp. 455–58.

Ehring, Douglas, "Dispositions and Functions: Cummins on Functional Analysis," *Erkenntnis* 23, pp. 243–50.

Finocchiaro, Maurice A., "Aspects of the Logic of History-of-Science Explanation," *Synthese* 62, pp. 429–54.

Grene, Marjorie, "Explanation and Evolution," in Robert S. Cohen and Marx Wartofsky, eds. *A Portrait of Twenty-five Years* (Dordrecht: D. Reidel Publishing Co.), pp. 177–97.

Hausman, David B., "The Explanation of Goal-Directed Behavior," *Synthese* 65, pp. 327–46.

Heil, John, "Rationality and Psychological Explanation," *Inquiry* 28, pp. 359–71.

☐ Humphreys, Paul, "Why Propensities Cannot Be Probabilities," *Philosophical Review* 94, pp. 557–70.

• Jobe, Evan K., "Explanation, Causality, and Counterfactuals," *Philosophy of Science* 52, pp. 357–89.

• Kitcher, Philip, "Two Approaches to Explanation," *Journal of Philosophy* 82, pp. 632–39.

——, van Fraassen, B., and Salmon, W. "Symposium on Wesley Salmon's *Scientific Explanation and the Causal Structure of the World,*" *Journal of Philosophy* 82, pp. 632–54.

☐ Mermin, N. David, "Is the Moon There When Nobody Looks? Reality and the Quantum Theory," *Physics Today* 38 (no. 4, April), pp. 38–47.

Olding, A., "Short on Teleology," *Analysis* 45, pp. 158–61.

Papineau, David, "Probabilities and Causes," *Journal of Philosophy* 82, pp. 57–74.

Pearce, David, and Veikko Rantala, "Approximative Explanation is Deductive Nomological," *Philosophy of Science* 52, pp. 126–40.

Prior, Elizabeth W., "What is Wrong with Etiological Accounts of Biological Function?" *Pacific Philosophical Quarterly* 66, pp. 310–28.

• Salmon, Wesley C., "Empiricism: The Key Question," in Nicholas Rescher, ed., *The Heritage of Logical Positivism* (Lanham, MD: University Press of America), pp. 1–21.

——(a), "Conflicting Conceptions of Scientific Explanation," *Journal of Philosophy* 82, pp. 651–54.

• Sober, Elliott, "A Plea for Pseudo-Processes," *Pacific Philosophical Quarterly* 66, pp. 303–9.

• van Fraassen, Bas C., "Salmon on Explanation," *Journal of Philosophy* 82, pp. 639–51.

Wilson, Fred, *Explanation, Causation, and Deduction.* Dordrecht: D. Reidel Publishing Co.

1986

Achinstein, Peter, "The Pragmatic Character of Explanation," in Peter D. Asquith and Philip Kitcher, eds., *PSA 1984*, vol. 2 (East Lansing, MI: Philosophy of Science Assn.), pp. 275–92.

☐ Barrow, John D., and Frank J. Tipler, *The Anthropic Cosmological Principle.* Oxford: Clarendon Press.

Bromberger, Sylvain, "On Pragmatic and Scientific Explanation: Comments on Achinstein's and Salmon's Papers," in Peter D. Asquith and Philip Kitcher, eds., *PSA 1984*, vol. 2 (East Lansing, MI: Philosophy of Science Assn.), pp. 306–25.

☐ Earman, John, *A Primer of Determinism.* Dordrecht: D. Reidel Publishing Co.

Forge, John, "The Instance Theory of Explanation," *Australasian Journal of Philosophy* 64, pp. 127–42.

Humphreys, Paul, "Review of *Scientific Explanation and the Causal Structure of the World* by Wesley C. Salmon," *Foundations of Physics* 16, 1211–16.

Kincaid, Harold, "Reduction, Explanation, and Individualism," *Philosophy of Science* 53, pp. 492–513.

Robinson, Joseph D., "Reduction, Explanation, and the Quests of Biological Research," *Philosophy of Science* 53, pp. 333–53.

Salmon, Wesley C., "Scientific Explanation: Three General Conceptions," in Peter D. Asquith and Philip Kitcher, eds., *PSA 1984*, vol. 2 (East Lansing, MI: Philosophy of Science Assn.), pp. 293–305.

Skarda, Christine A., "Explaining Behavior: Bringing the Brain Back in," *Inquiry* 29, pp. 187–202.

Smith, Quentin, "World Ensemble Explanations," *Pacific Philosophical Quarterly* 67, pp. 73–86.

Sober, Elliott, 'Explanatory Presupposition,' *Australasian Journal of Philosophy* 64, pp. 143–49.

1987

Bigelow, John, and Robert Pargetter, "Functions," *Journal of Philosophy* 84, pp. 181–96.

Fetzer, James H., "Critical Notice: Wesley Salmon's *Scientific Explanation and the Causal Structure of the World,*" *Philosophy of Science* 54, pp. 597–610.

Grimes, Thomas R., "Explanation and the Poverty of Pragmatics," *Erkenntnis* 27, pp. 79–92.

Irzik, Gurol, and Eric Meyer, "Causal Modeling: New Directions for Statistical Explanation," *Philosophy of Science* 54, pp. 495–514.

• Kitcher, Philip, and Wesley C. Salmon, "Van Fraassen on Explanation," *Journal of Philosophy* 84, pp. 315–30.

Levine, Joseph, "*The Nature of Psychological Explanation* by Robert Cummins," *Philosophical Review* 96, pp. 249–74.

Seager, William, "Credibility, Confirmation and Explanation," *British Journal for the Philosophy of Science* 38, pp. 301–17.

Sober, Elliott, "Explanation and Causation" (Review of Wesley Salmon, *Scientific Explanation and the Causal Structure of the World*), *British Journal for the Philosophy of Science* 38, pp. 243–57.

Torretti, Roberto, "Do Conjunctive Forks Always Point to a Common Cause?" *British Journal for the Philosophy of Science* 38, pp. 186–94.

The Fifth Decade

1988

Grünbaum, Adolf, and Wesley C. Salmon, eds., *The Limitations of Deductivism*. Berkeley & Los Angeles: University of California Press.

Koura, Antti, "An Approach to Why-Questions," *Synthese* 74, pp. 191–206.

Pitt, Joseph C., ed., *Theories of Explanation*. New York: Oxford University Press. [An anthology on scientific explanation.]

——(a), "Galileo, Rationality and Explanation," *Philosophy of Science* 55, pp. 87–103.

• Salmon, Wesley C., "Deductivism Visited and Revisited," in Grünbaum and Salmon, (1988).

——(a), "Intuitions: Good and Not-So-Good," in William Harper and Brian Skyrms, eds., *Causation, Cause, and Credence*, vol. I (Dordrecht: Kluwer Academic Publishers), pp. 51–71.

☐ Shimony, Abner, "The Reality of the Quantum World," *Scientific American* 258, no. 1 (Jan. 1988), pp. 46–53.

Temple, Dennis, "Discussion: The Contrast Theory of Why-Questions," *Philosophy of Science* 55, pp. 141–51.

1989

• Kitcher, Philip, and Wesley C. Salmon, eds., *Scientific Explanation*. Minneapolis: University of Minnesota Press.

Date Unknown

Humphreys, Paul, *The Chances of Explanation*. Princeton University Press.

van Fraassen, Bas C., *Laws and Symmetries*.

Index

Compiled by Charlotte A. Broome

220

Wesley C. Salmon is University Professor of Philosophy at the University of Pittsburgh. His principal earlier posts were at Brown University, Indiana University, and the University of Arizona, and he has been a visiting professor at several schools, including the Minnesota Center for Philosophy of Science. Salmon received his Ph.D. in philosophy from the University of California, Los Angeles. In 1977, he received a creative teaching award at the University of Arizona. Salmon's books include *Statistical Explanation and Statistical Relevance* (1971), *Space, Time, and Motion* (second edition, Minnesota, 1981), *Scientific Explanation and the Causal Structure of the World* (1984), and *Scientific Explanation* (co-edited by Philip Kitcher, Minnesota, 1989). He has served on the editorial board of several journals, including *Erkenntnis, Philosophy of Science*, and *Synthese*.